I0889958

HEALTH STANDARDS FOR LONG DURATION AND EXPLORATION SPACEFLIGHT

Ethics Principles, Responsibilities, and Decision Framework

Committee on Ethics Principles and Guidelines for Health Standards for Long Duration and Exploration Spaceflights

Board on Health Sciences Policy

Jeffrey Kahn, Catharyn T. Liverman, and Margaret A. McCoy, *Editors*

INSTITUTE OF MEDICINE
OF THE NATIONAL ACADEMIES

THE NATIONAL ACADEMIES PRESS
Washington, D.C.
www.nap.edu

THE NATIONAL ACADEMIES PRESS • 500 Fifth Street, NW • Washington, DC 20001

NOTICE: The project that is the subject of this report was approved by the Governing Board of the National Research Council, whose members are drawn from the councils of the National Academy of Sciences, the National Academy of Engineering, and the Institute of Medicine.

This project was supported by Contract NNH13CK14D between the National Academy of Sciences and the National Aeronautics and Space Administration. The views presented in this publication are those of the editors and attributing authors and do not necessarily reflect the view of the organizations or agencies that provided support for this project.

International Standard Book Number-13: 978-0-309-29657-1
International Standard Book Number-10: 0-309-29657-9

Additional copies of this report available for sale from the National Academies Press, 500 Fifth Street, NW, Keck 360, Washington, DC 20001; (800) 624-6242 or (202) 334-3313; http://www.nap.edu.

For more information about the Institute of Medicine, visit the IOM home page at: **www.iom.edu.**

Printed in the United States of America

The serpent has been a symbol of long life, healing, and knowledge among almost all cultures and religions since the beginning of recorded history. The serpent adopted as a logotype by the Institute of Medicine is a relief carving from ancient Greece, now held by the Staatliche Museen in Berlin.

Cover credit: All photos courtesy of the National Aeronautics and Space Administration

Suggested citation: IOM (Institute of Medicine). 2014. *Health standards for long duration and exploration spaceflight: Ethics principles, responsibilities, and decision framework.* Washington, DC: The National Academies Press.

"Knowing is not enough; we must apply.
Willing is not enough; we must do."

—Goethe

INSTITUTE OF MEDICINE

OF THE NATIONAL ACADEMIES

Advising the Nation. Improving Health.

THE NATIONAL ACADEMIES

Advisers to the Nation on Science, Engineering, and Medicine

The **National Academy of Sciences** is a private, nonprofit, self-perpetuating society of distinguished scholars engaged in scientific and engineering research, dedicated to the furtherance of science and technology and to their use for the general welfare. Upon the authority of the charter granted to it by the Congress in 1863, the Academy has a mandate that requires it to advise the federal government on scientific and technical matters. Dr. Ralph J. Cicerone is president of the National Academy of Sciences.

The **National Academy of Engineering** was established in 1964, under the charter of the National Academy of Sciences, as a parallel organization of outstanding engineers. It is autonomous in its administration and in the selection of its members, sharing with the National Academy of Sciences the responsibility for advising the federal government. The National Academy of Engineering also sponsors engineering programs aimed at meeting national needs, encourages education and research, and recognizes the superior achievements of engineers. Dr. C. D. Mote, Jr., is president of the National Academy of Engineering.

The **Institute of Medicine** was established in 1970 by the National Academy of Sciences to secure the services of eminent members of appropriate professions in the examination of policy matters pertaining to the health of the public. The Institute acts under the responsibility given to the National Academy of Sciences by its congressional charter to be an adviser to the federal government and, upon its own initiative, to identify issues of medical care, research, and education. Dr. Harvey V. Fineberg is president of the Institute of Medicine.

The **National Research Council** was organized by the National Academy of Sciences in 1916 to associate the broad community of science and technology with the Academy's purposes of furthering knowledge and advising the federal government. Functioning in accordance with general policies determined by the Academy, the Council has become the principal operating agency of both the National Academy of Sciences and the National Academy of Engineering in providing services to the government, the public, and the scientific and engineering communities. The Council is administered jointly by both Academies and the Institute of Medicine. Dr. Ralph J. Cicerone and Dr. C. D. Mote, Jr., are chair and vice chair, respectively, of the National Research Council.

www.national-academies.org

COMMITTEE ON ETHICS PRINCIPLES AND GUIDELINES FOR HEALTH STANDARDS FOR LONG DURATION AND EXPLORATION SPACEFLIGHTS

JEFFREY KAHN (*Chair*), Johns Hopkins University, Berman Institute of Bioethics, Baltimore, MD
NANCY CONRAD, The Conrad Foundation, Washington, DC
PETER F. DEMITRY, 4-D Enterprises, LLC, Haymarket, VA
BONNIE J. DUNBAR, University of Houston, TX
BARBARA J. EVANS, University of Houston Law Center, TX
BERNARD A. HARRIS, JR., Vesalius Ventures, Houston, TX
DAVID G. HOEL, Medical University of South Carolina
JONATHAN KIMMELMAN, McGill University, Montreal, Canada
ANNA C. MASTROIANNI, University of Washington School of Law, Seattle
LAWRENCE PALINKAS, University of Southern California
CAROL E. H. SCOTT-CONNER, University of Iowa Hospitals & Clinics
MICHAEL A. SILVERSTEIN, University of Washington School of Public Health
RONALD E. TURNER, ANSER (Analytic Services, Inc.), Falls Church, VA
R. LEONARD VANCE, Virginia Commonwealth University, Richmond
GREGORY R. WAGNER, National Institute for Occupational Safety and Health, Washington, DC

IOM Staff

CATHARYN T. LIVERMAN, Co-Study Director
MARGARET A. MCCOY, Co-Study Director
SARAH DOMNITZ, Associate Program Officer (*since December 2013*)
CLAIRE F. GIAMMARIA, Research Associate
JUDITH L. ESTEP, Program Associate
ASHNA KIBRIA, Research Associate (*since December 2013*)
R. BRIAN WOODBURY, Senior Program Assistant (*since December 2013*)
ANDREW M. POPE, Director, Board on Health Sciences Policy

Consultants

TRISH LEADER, Technical Writer
VICTORIA WEISFELD, Technical Writer

Reviewers

This report has been reviewed in draft form by individuals chosen for their diverse perspectives and technical expertise, in accordance with procedures approved by the National Research Council's Report Review Committee. The purpose of this independent review is to provide candid and critical comments that will assist the institution in making its published report as sound as possible and to ensure that the report meets institutional standards for objectivity, evidence, and responsiveness to the study charge. The review comments and draft manuscript remain confidential to protect the integrity of the deliberative process. We wish to thank the following individuals for their review of this report:

John R. Ball, American College of Physicians
Thomas F. Budinger, University of California, Berkeley
Naomi L. Gerber, George Mason University
Sven Ove Hansson, Royal Institute of Technology
Mae C. Jemison, BioSentient Corporation
Gloria Leon, University of Minnesota
Melissa A. McDiarmid, University of Maryland School of Medicine
Walter Schimmerling, East Carolina University
David H. Wegman, University of Massachusetts at Lowell
Laurie Zoloth, Northwestern University

Although the reviewers listed above have provided many constructive comments and suggestions, they did not see the final draft of the report before its release. The review of this report was overseen by **James F. Childress,** University of Virginia, and **Steve Fienberg,** Harvard University. Appointed by the Institute of Medicine, they were responsible for making certain that an independent examination of this report was carried out in accordance with institutional procedures and that all review comments were carefully considered. Responsibility for the final content of this report rests entirely with the authoring committee and the institution.

Preface

The prospect of long duration and exploration human spaceflight poses many significant challenges, not the least of which are the ethical issues raised by exposing astronauts to environments with uncertain and even unknown risks to their health, and excessive levels of known health risks. As a federally funded institution, National Aeronautics and Space Administration (NASA) recognizes that undertaking missions with such great risk and high levels of uncertainty raises challenges for decision making related to the health and safety of astronauts. In response to these concerns, the Institute of Medicine (IOM) convened a group with interdisciplinary expertise and a range of backgrounds to analyze the ethical issues faced when existing health standards will be exceeded or when health risks are uncertain or unknown.

This report represents the collective conclusions and recommendations of a diverse group of experts, each of whom brought their expertise and perspectives. The charge to the committee was clear. However it did not lend itself to the typical approach of collection of data, but instead largely relied on conceptual considerations and analysis. The resulting recommendations take the form of a collection of ethics principles and a framework for their application when faced with decisions about exceeding existing health standards or when standards do not exist. In recommending this collection of principles, the committee sought to identify and articulate the relevant moral rules that must be followed in the context of health standards for long duration and exploration spaceflight. The committee's considerations were informed by information from NASA, input from a range of stakeholders, and presentations from invited experts at public workshops. These principles and decision-making framework are not a checklist or recipe, but rather are intended to provide the parameters that must be satisfied for ethically acceptable health standards in these contexts. We hope they perform that function as NASA plans for future space exploration.

This report benefitted immensely from the dedicated, diligent, and skilled work of the IOM staff, including Meg McCoy, Cathy Liverman, Claire Giammaria, and Judy Estep; the committee gratefully acknowledges their tireless efforts.

Lastly, I want to thank my colleagues on the committee for their hard work and unflagging patience as we considered, and often reconsidered the ethical issues and challenges relevant to health standards for long duration and exploration spaceflight. It was a privilege to work with such a dedicated and insightful group.

Jeffrey Kahn, *Chair*
Committee on Ethics Principles and Guidelines
for Health Standards for Long Duration and
Exploration Spaceflights

Acknowledgments

The Institute of Medicine (IOM) Committee on Ethics Principles and Guidelines for Health Standards for Long Duration and Exploration Spaceflights would like to express its sincere gratitude to everyone who made this report possible. This report was informed by the contributions of many individuals who provided expertise, personal insights and perspectives, and data.

First, the committee would like to thank National Aeronautics and Space Administration (NASA) for its support of the study. The committee especially appreciates the leadership of Richard Williams, Chief Health and Medical Officer, NASA, who was instrumental in the development of this study and in leading the responses to a number of committee requests for information throughout the study. The committee appreciates input and assistance from many other current and former NASA staff members, including John Allen, Francis Cucinotta, Jeff Davis, Homayoon Dezfuli, Donna Good, Michael Griffin, David Liskowsky, Vincent Michaud, Victor Schneider, Marc Shepanek, Ed Simones, and Paul Wolpe. The insights gained from the NASA Astronaut Office, particularly from Michael Barratt, Robert Behnken, Shannon Walker, and Peggy Whitson, provided important context for the committee's work. The thorough and considered input provided to all of the committee's queries was much appreciated.

The committee held two public workshops in May and July 2013 and gained valuable insights from the substantive presentations provided by the speakers (see Appendix A) and participants.

The committee greatly benefited from the work of the IOM study staff team: Cathy Liverman and Meg McCoy co-directed the study; Claire Giammaria provided research support; and Judy Estep skillfully managed all logistical and administrative aspects of the study. Our thanks to Andrew Pope for his leadership. Additionally, Sarah Domnitz, Ashna Kibria, and Brian Woodbury were instrumental in finalizing the report.

The committee is also grateful to Trish Leader and Vicki Weisfeld of NEW Associates, LLC, for their assistance in editing the report and to Laura Penny for copyediting the report.

Finally, the committee would like to express its thanks to the IOM and the National Research Council staff members who worked behind the scenes to ensure a seamless study process and successful production and dissemination of this report: Anton Bandy, Clyde Behney, Daniel Bethea, Laura DeStefano, Chelsea Frakes, Molly Galvin, Janice Mehler, Donna Randall, Lora Taylor, Erika Vijh, and Jennifer Walsh.

Contents

Summary

Since its inception, the U.S. human spaceflight program has grown from launching a single man into orbit to an ongoing space presence aboard the International Space Station (ISS) with partners from many nations. Human spaceflight inherently involves a high degree of risk during all phases of any space mission, including terrestrial training and vehicle testing, launch, inflight during the mission, and landing. Health risks during long duration and exploration spaceflights include short-term health consequences (e.g., nausea or fatigue from acute radiation exposure during a solar storm, injury, blurred vision), as well as long-term health consequences that arise or continue months or years after a flight (e.g., radiation-induced cancers, loss of bone mass). Long duration and exploration spaceflights (including extended stays on the ISS or exploration missions to an asteroid or Mars) will likely expose crews to levels of known risks beyond those allowed by current health standards, as well as to a wide range of risks that are poorly characterized, uncertain, and perhaps unforeseeable.

The National Aeronautics and Space Administration (NASA) asked the Institute of Medicine (IOM) to develop an ethics framework and to identify principles to guide decision making about health standards for long duration and exploration class missions "when existing health standards cannot be fully met" or adequate standards cannot be developed based on existing evidence (see Box S-1). The committee was not asked to review NASA's risk management processes.

BOX S-1
Statement of Task

The National Aeronautics and Space Administration (NASA) is in the process of planning for exploration class missions of long duration and beyond low Earth orbit (LEO). An ad hoc committee of the Institute of Medicine (IOM) will conduct a study to examine policy and ethical issues relevant to crew health standards for these missions. The committee would consider the application of existing health standards and the potential development of a new set of standards for missions beyond LEO. These standards would address potential hazardous exposures and working conditions that are uncertain, unknown, or that go beyond current NASA risk limits. NASA is looking, in particular, for a framework of ethical and policy principles that can help guide decision making associated with implementing health standards for exploration class space missions when existing standards cannot be fully met, or the level of knowledge of a given condition is sufficiently limited that an adequate standard cannot be developed, for the mission. As part of its deliberations, the committee will consider and respond to options proposed by NASA, as well as offer options of its own. Given the long-term importance of this task for NASA, the committee's report will fully detail supporting rationale for all recommendations.

Questions to be considered include

1. What factors should be considered in the implementation of current health standards (which are defined in Volumes 1 and 2 of *NASA Space Flight Human System Standards*) in exploration class missions and the development of exploration class health standards if necessary?
   a. What ethical considerations are involved in developing and implementing health and safety standards for manned space exploration when the exposures and risks are uncertain, unknown, and/or when exposures and risks might exceed current standards?
   b. What standards of informed consent regarding the health risks of the mission are appropriate, and what are the ethical limits of informed consent processes in these circumstances? What principles should be applied (when relevant) to communicating the uncertainty regarding health risks?
   c. What are appropriate modifiers for standards for protecting individuals when there is an incomplete understanding and knowledge of the potential risks or hazardous exposures, or when exposures and risks may exceed current standards?
   d. Should all astronauts and spaceflight crew members be protected to the same risk level or should potential individual differences be considered? Would one standard be sufficient for the entire spaceflight crew or do known or unknown differences in risk need to be addressed to provide a uniform level of protection?

> 2. Are there models or examples of other situations with unknown health risks (or risks that could exceed current standards) that could inform NASA policy and, if so, how?

To respond to its statement of task, the IOM convened the Committee on Ethics Principles and Guidelines for Health Standards for Long Duration and Exploration Spaceflights. This report provides the committee's recommendations on ethics principles, responsibilities, and a decision framework to guide decisions about the application of health standards to long duration and exploration spaceflights. The recommendations include many components that NASA has considered or is already implementing in its work on health standards for spaceflight. However, the adoption of the recommended ethics principles and responsibilities and decision framework not only introduces new processes and responsibilities but also alters the context in which existing components operate.

HEALTH STANDARDS FOR HUMAN SPACEFLIGHT

The use of health standards to protect individuals engaged in specific types of work or activities is not unique to space exploration. Health standards are used throughout occupational settings to protect workers and guide design, research, and engineering efforts, among other purposes. The committee examined various high-risk occupations and activities and identified some common factors (such as types and severity of risk; types and distribution of potential risks and benefits; activity purpose; and the presence of independent oversight) that aided the committee's deliberations. However, the committee was not able to identify existing ethics frameworks or decision-making models for specific occupations that were directly and wholly applicable to decisions about health standards for long duration and exploration spaceflight.

As the federal agency with institutional responsibilities to conduct and evaluate a wide range of spaceflight missions, NASA addresses health risks associated with spaceflight using a number of strategies, including engineering, design, mission planning, basic and clinical research, surveillance and medical monitoring, preventive and treatment countermeasures, and health standards. NASA's crew health standards consider three categories of preflight, inflight, and postflight health issues: fitness-for-duty standards, permissible exposure limits, and permissible outcome limits. Additionally, NASA has health standards relevant to human factors, habitability, and environmental health.

If a human spaceflight mission cannot meet current health standards, or if inadequate information exists to revise a health standard, the options, as identified and examined by the committee, would be to: (1) liberalize the NASA health standards, (2) establish more permissive "long duration and exploration health standards," or (3) grant an exception to the standard in order to conduct these missions before new protective technologies and strategies are available or additional data are acquired which may allow revision of the standard. The committee found the first two options to be unacceptable when evaluated against the ethics principles and responsibilities described in this report.

The committee finds relaxing (or liberalizing) current health standards to allow for specific long duration and exploration missions to be ethically unacceptable. Current NASA policy outlines the administrative processes and levels of approval required to initiate a new health standard or revise a current health standard. NASA's health standards "are based on the best available scientific and clinical evidence, as well as operational experience." Moreover, review of health standards occurs every 5 years or can be triggered any time that new research data or clinical observations indicate an update may be necessary. Modifying health standards outside of this established process merely to permit long duration and exploration missions would be arbitrary.

The second option maintains current health standards for all missions except long duration and exploration spaceflight, which would have a separate set of health standards. This approach presumably would entail setting a more permissive ceiling on allowable risk for long duration and exploration missions under conditions in which existing evidence and knowledge make it nearly impossible to quantify those ceilings. The committee found this approach wanting, lacking a clear and compelling justification for why acceptable risks and levels of uncertainty should be greater for long duration and exploration missions than other human spaceflight missions.

Having excluded the options of modifying existing standards or creating a separate set of standards, the committee concludes that the only ethically acceptable option that could allow for increased risk exposures in the context of long duration and exploration spaceflights is granting an exception to existing health standards. The committee believes that exceptions to health standards should be considered on a mission-by-mission basis and used in very limited circumstances following the ethics-based framework recommended.

When making decisions about whether to allow risk levels that do not meet NASA's current health standards, it is important that current health standards reflect the most relevant and up-to-date evidence about spaceflight-related risks to human health and safety. Although NASA has formal policies for health standard initiation and revision, the committee believes additional information about decision criteria, process, and application of ethics principles should be available to the public.

> **Recommendation 1:** ***Expand on the Policies for Initiating and Revising Health Standards***
> **NASA should ensure that its policies regarding health standards detail the conditions or circumstances (and relevant priorities) that initiate development or revision of health standards and explicitly indicate how these policies are fully consistent with the set of ethics principles outlined in this report.**

ETHICS PRINCIPLES

Over the course of centuries, philosophers have debated the relative merits and shortcomings of the major theories of western moral and political philosophy: utilitarianism, duty-based approaches, virtue-based theories, and others. One successful approach to avoiding the need for a single ethical theory is to focus on mid-level principles rather than the theory to which they belong or from which they are derived. This approach has proven to be particularly successful when used by expert committees or commissions made up of individuals with diverse commitments in an effort to find common ground on how best to approach challenging ethical issues in the context of public policy. The committee adopted a principle-based approach in recognition of its accessibility and applicability for its task.

> **Recommendation 2:** ***Apply Ethics Principles to Health Standards Development and Implementation***
> **NASA should apply the following ethics principles in the development and implementation of its health standards for decisions regarding long duration and exploration spaceflights:**

- ***Avoid harm***—the principle includes the duty to prevent harm, exercise caution, and remove or mitigate harms that occur. Thus, NASA should exhaust all feasible measures to minimize the risks to astronauts from long duration and exploration spaceflights, including addressing uncertainties through approaches to risk prevention and mitigation that incorporate safety margins and include mechanisms for continuous learning that allow for incremental approaches to risk acceptance.
- ***Beneficence***—the principle to provide benefit to others. NASA should consider in its decision making the potential benefits of a specific mission, including its scientific and technological importance, as well as its potential beneficiaries including current and future astronauts and members of society at large.
- ***Favorable balance of risk and benefit***—the principle to seek both a favorable and acceptable balance between the risk of harm and potential for benefit. In authorizing long duration and exploration activities and in approving particular missions, NASA should systematically assess risks and benefits and the uncertainties attached to each, drawing on the totality of available scientific evidence, and ensuring that benefits sufficiently outweigh risks.
- ***Respect for autonomy***—the principle to ensure that individuals have both the right to self-determination and processes in place to exercise that right. NASA should ensure that astronauts are able to exercise voluntariness to the extent possible in personal decision making regarding participation in proposed missions, that they have all available information regarding the risks and benefits of the proposed mission, and that they continue to be apprised of any updates to risk and benefit information throughout the mission.
- ***Fairness***—the principle requires that equals be treated equally, that burdens and benefits be distributed fairly, and that fair processes be created and followed. NASA's decision making surrounding missions should explicitly address fairness, including the distribution of the risks and benefits of the mission, crew selection, and protections for astronauts after missions.

- ***Fidelity***—the principle recognizes that individual sacrifices made for the benefit of society may give rise to societal duties in return. Given the risks that astronauts accept in participating in hazardous missions, NASA should respect the mutuality of obligations and ensure health care and protection for astronauts not only during the mission but after return, including provision of lifetime health care for astronauts.

ETHICS-BASED RESPONSIBILITIES

Acting on ethics principles creates specific responsibilities that apply to individuals and organizations engaged in particular activities. NASA, as an employer, a federal agency responsible for innovation and exploration, a research sponsor, and an international partner, has moral obligations to formally recognize and act on responsibilities that logically flow from the ethics principles outlined in Recommendation 2. Specifically, the ethics principles need to be incorporated into decisions about whether risks in excess of those allowed under current health standards are acceptable and, if so, what conditions must be satisfied to engage in ethically acceptable long duration and exploration class missions. These decisions are further influenced by the limitations imposed by costs, time lines, and technological feasibility.

Recommendation 3: *Implement Ethics Responsibilities*
NASA should adopt policies or processes that formally recognize the following ethics responsibilities related to health standards for long duration and exploration spaceflights:

- **Fully inform astronauts about the risks of long duration and exploration spaceflights and make certain that the *informed decision-making process* is adequate and appropriate.**
- **Adhere to a *continuous learning strategy* (including health surveillance and data collection) to ensure that health standards evolve and improve over time and are informed by data gained before, during, and after long duration and exploration spaceflights, as well as from other relevant sources.**

- **Solicit *independent advice* about any decision to allow any specific mission that fails to meet NASA health standards or any decision to modify health standards.**
- **Communicate with all relevant stakeholders (such as astronauts and the public at large) the rationale for, and possible impacts (including harm type, severity, and probability estimates) related to any decision about health standards in a *procedurally transparent, fair and timely manner*, providing adequate opportunity for public engagement.**
- **Provide *equality of opportunity* for participation in long duration and exploration spaceflights to the fullest extent possible. For example, fairness in crew selection means that NASA should accept some group differences in population risk in order to create equality of opportunity to participate in missions, and accommodate individual variance from population-based risk estimates to the extent that individual differences do not jeopardize mission operations.**
- **Provide *preventive long-term health screening and surveillance* of astronauts and *lifetime health care* to protect their health, support ongoing evaluation of health standards, improve mission safety, and reduce risks for current and future astronauts.**
- **Develop and apply policies that appropriately and sufficiently *protect the privacy and confidentiality* of astronaut health data.**

AN ETHICS-BASED DECISION FRAMEWORK

In assessments of long duration and exploration spaceflights that are unlikely to meet current health standards, NASA can facilitate its planning by structuring decision making in a three-step process (see Box S-2). The first and broadest decision (Level 1) is whether, and under what conditions, any missions that are unlikely to meet current health standards are ethically acceptable. If NASA decides that missions that fail to meet existing health standards are not acceptable, then such missions must be deferred until new knowledge about risk or uncertainties or risk mitigation strategies are available.

If NASA decides that exceptions to existing health standards are ethically acceptable, NASA must then decide what process and criteria should be applied to determine whether a specific mission should be granted an exception. Based on the ethics principles identified, criteria for reviewing exception requests could include requirements that the proposed mission

- be expected to have exceptionally great social value;
- have great time urgency;
- have expected benefits that would be widely shared;
- be justified over alternate approaches to meeting the mission's objectives;
- establish that existing health and safety standards cannot be met;
- be committed to minimizing harm and continuous learning;
- have a rigorous consent process to ensure that astronauts are fully informed about risks and unknowns, meet standards of informed decision making, and are making a voluntary decision; and
- provide health care and health monitoring for astronauts before, during, and after flight and for the astronaut's lifetime.

An ethically acceptable process to grant exceptions would also be evidence-based, transparent, and solicit independent advice. The committee emphasizes that exceptions should only be granted in rare circumstances and should increase the responsibilities for NASA and society.

The second decision level (Level 2) is whether a *specific, contemplated mission* that is unlikely to meet current health standards is ethically acceptable. This requires evaluating whether the mission fulfills the conditions articulated in Level 1. Whereas Level 1 could be a one-time decision by NASA that affects all long duration and exploration missions,[1] Level 2 decisions would be made on a mission-by-mission basis.

Once a specific mission is deemed ethically acceptable, the third level (Level 3) of decision making involves identification of the crew for the mission given the mission's requisite skills and expertise, the astronauts' health susceptibilities and personal risk factors (if known), and an individual astronaut's informed decision to participate. Level 3 includes decisions by both NASA and individual astronauts.

---

[1]It is important to note that this decision may be revisited by NASA but should adhere to the same ethics-based decision framework to justify any alterations.

Recommendation 4: *Adopt an Ethics-Based Decision Framework*
NASA should apply the relevant ethics principles and fulfill the concomitant responsibilities through a three-level, ethics-based decision framework that examines

- **Level 1: Decisions about allowing risk to astronaut health and safety in excess of that permitted by health standards,**
- **Level 2: Decisions about undertaking specific missions, and**
- **Level 3: Decisions concerning individual astronaut participation and crew composition.**

BOX S-2
Decision Framework

Level 1: Decisions About Missions That Fail to Meet Health Standards

Decision points:

- Should NASA conduct space missions that will (a) fail to meet health standards, (b) involve significant risks where there are no applicable standards, and/or (c) involve such great uncertainty that NASA cannot exclude the possibility of a or b?
- If so, what criteria should be used to determine whether exceptions for specific missions should be allowed?

Ethics principles and applications: Avoid harm, beneficence, acceptable risk-benefit balance, fidelity, transparency of decision making, commitment to continuous learning, procedural fairness of decision making

Examples of ethics responsibilities:

- Ensure that all feasible means will be taken to reduce astronaut risks to the lowest achievable levels
- Examine all approaches to minimizing risk including alternate approaches to meeting the mission's objectives
- Assess and communicate the benefits
- Determine and communicate the time urgency to conduct the mission
- Thoroughly monitor and conduct research on health impacts during and after spaceflight to inform current and future missions
- Commit to the future health of current and future astronauts by ensuring access to health care, longitudinal follow-up, and preventive screenings

Level 2: Decisions About Specific Missions

Decision point: Given authorization for missions that will likely fail to meet existing health standards, is a specific long duration and/or exploration mission ethically acceptable?

Ethics principles and applications: Avoid harm, acceptable risk-benefit balance, transparency of decision making, commitment to continuous learning, procedural fairness of decision making

Examples of ethics responsibilities:

- Adherence to criteria that are established and transparent
- Share risk escalation decisions and strategies
- Continue independent input to standards development and refinement
- Implement a robust program of occupational health monitoring and data collection during and after the mission
- Demonstrate that standards cannot be met despite having taken all feasible measures to reduce exposures to the lowest achievable level

Level 3: Decisions About Crew Selection and Individual Astronaut Participation

Decision point: What factors should be considered as NASA and individual astronauts make informed decisions about crew selection and individual astronaut participation for a given mission?

Ethics principles and applications: Informed decision making by astronauts, fairness, avoid harm, risk minimization (including risk to other crew members), commitment to continuous learning while protecting privacy and confidentiality

Examples of ethics responsibilities:

- Thorough sharing of risk data with astronauts
- Transparent and fair processes and policies on decision making
- Astronaut responsibilities to participate in data collection and health monitoring during and after spaceflights to inform current and future crews
- Selecting crew members in a manner that ensures fairness among groups and considers risk susceptibilities in general and for individuals in a way that allows inclusion, individual decision making within a range of risk that is prudent for the welfare of all astronauts during the mission

CONCLUDING REMARKS

From its inception, human spaceflight has pushed the boundaries of acceptable health and safety risks for astronauts. NASA, its international partners, and commercial space ventures will continue to face complex ethical decisions about risk acceptability as technologies improve and longer and more distant spaceflights become feasible. The ethics framework outlined in this report builds on the work of NASA and others and identifies a set of recommendations for ethically assessing and responding to the challenges associated with health standards for long duration and exploration spaceflight. Establishing and maintaining a firmly grounded ethics framework for this inherently risky activity is essential to guide NASA's decisions today and to create a strong foundation for decisions about future challenges and opportunities.

1

Introduction

Every age has its dreams, its symbols of romance. Past generations were moved by the graceful power of the great windjammers, by the distant whistle of locomotives pounding through the night, by the caravans leaving on the Golden Road to Samarkand, by quinqueremes of Nineveh from distant Ophir....Our grandchildren will likewise have their inspiration—among the equatorial stars. They will be able to look up at the night sky and watch the stately procession of the Ports of Earth—the strange new harbors where the ships of space make their planetfalls and their departures.

—Sir Arthur C. Clarke, The Promise of Space, *1968, p. 125*

From the dawn of oral and recorded history, as enshrined in the words of Homer and Herodotus, humans have pursued the unknown. Exploration undertaken in the name of adventure and scientific innovations, new trade markets, geopolitical dominance, and the advancement of human civilization has led to the discovery of distant lands and new ideas that challenge accepted beliefs and perceived limitations. Naval and land exploration have connected remote cultures and civilizations, generated new sources of wealth, expanded scientific and technological knowledge and capabilities, and promoted an exchange of ideas that has revolutionized long-held beliefs about the world.

The last half of the 20th century welcomed an era of great potential for discovery. In 1961, Soviet cosmonaut Yuri Gagarin became the first human in space (NASA, 2014c), once again expanding humanity's perception of the possible. Later that year, U.S. astronaut Alan Shepard became the first American to fly into space and, in 1962, John Glenn made history as the first U.S. astronaut to orbit Earth (Garber and Launius, 2005). Since those monumental firsts, astronauts have walked on the Moon; launched, retrieved, and repaired various satellites (e.g., the Hubble Space Telescope); conducted thousands of scientific and engineering research experiments; and assembled and resupplied the International Space Station (ISS) (NRC, 2011). Investment in space exploration has led to cutting-edge aerospace technology, weather satellites and climate modeling, communication and navigation satellites, new and transformational medical technologies, and a greater understanding of the effects of space on human physiology (Dick and Cowing, 2004). However, these achievements have come at great cost.

Exploration, especially in relatively new environments, involves some known risks and a wider range of unknown risk, which jeopardize the safety of explorers and their teams as well as the mission. For example, during the Age of Discovery (15th to 17th centuries), unprecedented naval explorations often required extended voyages during which crews experienced hardship, deprivation, and disease (Harrison, 2013). The dynamic and unpredictable conditions of extreme environments, often complicated by a lack of available resources and equipment failures, have taken countless lives over many centuries (e.g., Solomon, 2001; Davis, 2011). In the context of space exploration, more than 20 U.S. astronauts have died during, and in preparation for, spaceflight (NASA, 2014a). But past failures, as well as successes, provide opportunities to critically examine, anticipate, and mitigate similar risks to future explorers.

As modern explorers continue to push the limits of human endurance in even more remote and inhospitable environments, risks to individuals and other interested parties will also change, raising new questions about risk tolerance limits and thresholds of acceptable risk for both individuals and society. Responsible decision making will call for a careful articulation of the values that shape the balance our society strikes between the quest for discovery and avoidance of harm.

SCOPE OF WORK

The United States formally launched efforts to engage in "aeronautical and space activities" in 1958, when Congress passed the National Aeronautics and Space Act, establishing the civilian-controlled NASA.[1] NASA's charge expanded the scope of federally supported aeronautical research,[2] to include "the problems of flight *within and outside* the Earth's atmosphere," excepting activities related to weapon systems, national security, and military activities.[3] Specifically, Congress directed NASA to engage in activities "devoted to peaceful purposes for the benefit of all mankind" with specific objectives, including

(1) The expansion of human knowledge of phenomena in the atmosphere and space;
(2) The improvement of the usefulness, performance, speed, safety, and efficiency of aeronautical and space vehicles;
(3) The development and operation of vehicles capable of carrying instruments, equipment, supplies, and living organisms through space;
(4) The establishment of long-range studies of the potential benefits to be gained from, the opportunities for, and the problems involved in the utilization of aeronautical and space activities for peaceful and scientific purposes;
(5) The preservation of the role of the United States as a leader in aeronautical and space science and technology and in the application thereof to the conduct of peaceful activities within and outside the atmosphere;
(6) The making available to agencies directly concerned with the national defense of discoveries that have military value or significance, and the furnishing by such agencies, to the civilian agency established to direct and control nonmilitary aeronautical and space activities, of information as to discoveries which have value or significance to that agency;

---

[1]The National Aeronautics and Space Act of 1958, P.L. 85-568 (July 29, 1958).

[2]The National Advisory Committee on Aeronautics, NASA's predecessor, was founded to "supervise and direct the scientific study of the problems of flight, with a view to their practical solution, and to determine the problems which should be experimentally attacked, and to discuss their solution and their application to practical questions" (The Naval Appropriations Act of 1916, P.L. 63-271, [March 3, 1915]).

[3]The National Aeronautics and Space Act of 1958, P.L. 85-568 (July 29, 1958).

(7) Cooperation by the United States with other nations and groups of nations in work done pursuant to this Act and in the peaceful application of the results, thereof; and

(8) The most effective utilization of the scientific and engineering resources of the United States, with close cooperation among all interested agencies of the United States in order to avoid unnecessary duplication of effort, facilities, and equipment.[4]

These objectives have played an integral role in shaping NASA's history and vision, and are important considerations when exploring the ethics defining health standards in human spaceflight.[5] NASA is charged with designing and implementing space missions, which pose high risks to human health and safety, and with risk assessment and management. This broad governance structure provides opportunities to design feedback loops that can translate data and experience into improved risk management strategies, but it can also lead to perceived and actual conflicts of interest that have important ethical implications, as discussed later in this report.

Moreover, important shifts in human space exploration continue to redefine the sphere in which space travel occurs. NASA's human spaceflight program has grown from a one-man launch to an ongoing space presence aboard the ISS, involving men and women from more than a dozen nations (NASA, 2013c). The Moon, once a short-term destination, is now considered by some as a possible permanent base for human activity (Duke et al., 1985; Angelo, 2009; Duke, 2013; NASA, 2014b). The U.S. government has articulated visions that include a foreseeable human presence on the Moon (once again) and on Mars' surface.[6,7] Long duration and exploration spaceflights involving humans appear to be on the immediate horizon (NASA, 2004; Bolden, 2013; Norwood and Tahu,

[4]Congress has since expanded NASA's objectives to include additional responsibilities, such as the "preservation of the United States preeminent position in aeronautics and space through research and technology development related to associated manufacturing processes" (National and Commercial Space Programs, P.L. 111-314 [December 18, 2010]).

[5]This report does not review the history of NASA or the U.S. space program, but numerous texts are available (e.g., Garber and Launius, 2005; Dick, 2008, 2010; Logsdon and Launius, 2008).

[6]National Aeronautics and Space Administration Authorization Act of 2005, P.L. 109-155 (December 30, 2005).

[7]National Aeronautics and Space Administration Authorization Act of 2008, P.L. 110-422 (October 15, 2008).

2013), and discussions about sending humans to an asteroid and Mars have stimulated debate about the fiscal and ethical limits of spaceflight.

Contemporary space exploration is also an increasingly cooperative effort. Commercial and international collaborations propel efforts to institute a sustained human presence deeper into the solar system. Multiple companies are developing their own space vehicles and business plans (NASA, 2013b). With the retirement of the Space Shuttle Program, NASA has had to rely on other governments and space agencies, and will, in the future, rely on the commercial sector to provide low Earth orbit (LEO)[8] transportation for its astronauts to NASA-supported facilities in space (NASA, 2013d). Increased collaboration has also led to challenges associated with harmonizing regulations and policies, including those that govern health risks to astronauts.

Within NASA, the Human Exploration and Operations Mission Directorate provides "leadership and management of NASA space operations related to human exploration in and beyond [LEO]" as well as "NASA space operations related to Launch Services, Space Transportation, and Space Communications in support of both human and robotic exploration programs" (NASA, 2013a). Recognizing that the acceptance of increased levels of risk brings with it a profound responsibility to minimize and mitigate risks to the greatest extent feasible (Bolden, 2013), NASA has implemented a comprehensive risk management policy, as described in Chapter 2, which informs the conception, development, and execution of NASA programs and projects (NASA, 2011a). This policy includes consideration of the risks and benefits of specific human spaceflight missions.

One of NASA's primary risk management activities includes developing and enforcing health standards to provide a "healthy and safe environment for crewmembers, and to provide health and medical programs for crewmembers during all phases of space flight" (NASA, 2007, p. 8). The health standards examined in this report are specified in Volumes 1 and 2 of the *NASA Spaceflight Human System Standards* (NASA, 2007, 2011b). The committee uses the shorter term "health standards" to refer to this set of standards throughout this report. NASA's health standards include preflight, inflight, and postflight health issues and fall into three categories: fitness-for-duty standards, space-permissible exposure limits, and permissible outcome limits (NASA, 2007).

---

[8]LEO is the orbit above Earth's surface at altitudes between 100 miles (160 km) and approximately 1,240 miles (2,000 km) (National Geographic, 2014).

The degree to which NASA astronauts and facilities can take part in collaborative space-exploration activities is governed, in part, by the health standards adopted to protect individual astronauts from the adverse health effects of spaceflight. Long duration and exploration spaceflights will involve increased exposures of crews to known risks, as well as to a wide range of risks that are poorly characterized and, perhaps, unforeseeable. "Despite superficial similarities to other space missions and analogues, the extended durations and astronomical distances involved in lunar and [M]artian missions will make these activities far more difficult and dangerous" (Stuster, 1996, p. 3). As NASA and Congress continue to discuss the next generation of NASA's missions and the U.S. role in international space efforts, it is important to understand the ethical factors that drive decision making surrounding health standards and mission design for NASA activities.

Committee Charge

Despite the known and unknown health risks, some individuals would undoubtedly agree to serve on exploration and long duration spaceflights. According to the Institute of Medicine (IOM) report *Safe Passage: Astronaut Care for Exploration Missions*, "the opportunity to travel in space is highly desired by those in the astronaut corps[, and] although long-duration space missions will involve substantial hazards, there are likely to be many who will gladly accept the risk in exchange for the unique opportunity to leave the bounds of Earth's orbit" (IOM, 2001, p. 179). Given uncertainties about health and safety risks facing astronauts during extended stays on the ISS or on exploration missions to an asteroid or Mars, NASA asked the IOM to examine the ethics and policy principles that should guide decision making about health standards for long duration and exploration class missions "when existing health standards cannot be fully met" or adequate standards cannot be developed based on existing evidence (see Box 1-1). Based on this language from its task, the committee decided to use the phrase "fails to meet" or variants of this phrase to indicate when conditions or uncertainties related to spaceflights either may or will subject astronauts to higher levels of risk than those allowed by existing health standards.

To respond to the statement of task (see Box 1-1), the IOM convened a 15-member Committee on Ethics Principles and Guidelines for Health Standards for Long Duration and Exploration Spaceflights ("the committee"), which included experts in ethics, spaceflight and aerospace medicine, occupational health, risk assessment, law, behavioral health, and health standards. In essence, the committee's statement of task asks which ethics principles need to be considered in questions about acceptable risks to humans under highly uncertain conditions, and how these principles should be applied to decisions about long duration or exploration spaceflights sponsored by the federal government.

BOX 1-1
Statement of Task

The National Aeronautics and Space Administration (NASA) is in the process of planning for exploration class missions of long duration and beyond low Earth orbit (LEO). An ad hoc committee of the Institute of Medicine (IOM) will conduct a study to examine policy and ethical issues relevant to crew health standards for these missions. The committee would consider the application of existing health standards and the potential development of a new set of standards for missions beyond LEO. These standards would address potential hazardous exposures and working conditions that are uncertain, unknown, or that go beyond current NASA risk limits. NASA is looking in particular for a framework of ethical and policy principles that can help guide decision making associated with implementing health standards for exploration class space missions when existing standards cannot be fully met, or the level of knowledge of a given condition is sufficiently limited that an adequate standard cannot be developed, for the mission. As part of its deliberations, the committee will consider and respond to options proposed by NASA, as well as offer options of its own. Given the long-term importance of this task for NASA, the committee's report will fully detail supporting rationale for all recommendations.

Questions to be considered include

1. What factors should be considered in the implementation of current health standards (which are defined in Volumes 1 and 2 of *NASA Space Flight Human System Standards*) in exploration class missions and the development of exploration class health standards if necessary?
   a. What ethical considerations are involved in developing and implementing health and safety standards for manned space exploration when the exposures and risks are uncertain, unknown, and/or when exposures and risks might exceed current standards?
   b. What standards of informed consent regarding the health risks of the mission are appropriate, and what are the ethical limits of informed consent processes in these circumstances? What principles should be applied (when relevant) to communicating the uncertainty regarding health risks?

c. What are appropriate modifiers for standards for protecting individuals when there is an incomplete understanding and knowledge of the potential risks or hazardous exposures, or when exposures and risks may exceed current standards?
d. Should all astronauts and spaceflight crew members be protected to the same risk level or should potential individual differences be considered? Would one standard be sufficient for the entire spaceflight crew or do known or unknown differences in risk need to be addressed to provide a uniform level of protection?

2. Are there models or examples of other situations with unknown health risks (or risks that could exceed current standards) that could inform NASA policy and, if so, how?

Clear definitions of several key terms are essential to understand the scope of the committee's task. NASA has defined "exploration" missions as "any activity outside of the radiation protection of the near-Earth magnetic field (i.e., Moon, asteroid, Mars)" and "long duration missions" as those extending beyond 30 days.[9] However, to be consistent with its statement of task, the committee generally refers to exploration class missions as those beyond LEO. Missions to areas beyond LEO will involve radiation exposures above those consistently encountered on the ISS. In the context of the committee's charge, these areas are conceptually important because they not only help define the directive of a mission (i.e., to reach a particular destination in space), but they also define the parameters of risk exposure and the resulting impact of current health standards on a specific mission.

This report focuses on the benefits of human spaceflight only to the extent that it is relevant to the identification of ethics principles and the development of a decision framework related to health standards for long duration and exploration spaceflights (see Chapters 5 and 6). The committee was not tasked with drawing conclusions or making recommendations about the value or advisability of future human spaceflight, the prioritization of NASA activities, or the appropriate level of funding NASA should receive to support human spaceflight activities. Moreover, this report does not examine the complexities of quantitative risk assessment and risk-benefit analyses, processes that should influence NASA's decisions about long duration and exploration spaceflight. Instead, the

[9]Personal communication, Richard Williams, NASA, November 19, 2013.

scope of this report is limited to identifying an ethics framework upon which other data and analyses should be evaluated.

In addition to reviewing NASA documents and relevant literature, the committee held two public workshops (see Appendix A for the workshop agendas), one public teleconference, and six closed-session meetings over the course of 10 months to inform its deliberations. At the public workshops and open sessions, the committee solicited input about the challenges, goals, and consequences associated with health standards for long duration and exploration class missions. The committee heard from NASA experts involved in risk assessment and management for the agency, as well as a number of past and current astronauts. Throughout this study, NASA leadership and staff provided substantial cooperation, support, and responsiveness as the committee sought additional information for its deliberations. The committee also solicited input from experts on the ethics of risk acceptance, decision making in the context of high-risk occupations, and federal agency health standard implementation and enforcement.

Related Work of the National Academies

The National Research Council (NRC) and the IOM have been involved in providing advice to NASA about space travel and human space exploration since the late 1950s (NRC, 2013a) and have produced numerous studies, such as *Safe Passage: Astronaut Care for Exploration Missions* (IOM, 2001); *Review of NASA's Longitudinal Study of Astronaut Health* (IOM, 2004); *Review of NASA's Human Research Program Evidence Books: A Letter Report* (IOM, 2008); and *Managing Space Radiation Risk in the New Era of Space Exploration* (NRC, 2008). The IOM and the NRC have also addressed risk acceptability and decision making in many other contexts (NRC, 1996, 2009), including environmental settings (IOM, 2013). These reports have helped inform the approach recommended in the present report.

The NRC is currently conducting a study examining the "long-term goals, core capabilities, and direction of the U.S. human spaceflight program" to "provide findings, rationale, prioritized recommendations, and decision rules that could enable and guide future planning for [sustainable] U.S. human spaceflight exploration" (NRC, 2013b). As part of the work to address its charge, the NRC committee will consider the rationales for human space exploration. The committee will also describe the value of NASA's human spaceflight activities in the context of national goals (NRC, 2013b) and international partnerships.

REPORT STRUCTURE

This report is written for a broad audience that may not be familiar with the technical aspects of NASA's health standards or the intricacies of ethics. Chapters 2 and 3 of the report detail NASA's risk management process in the context of current health standards and provide a few examples of relevant health risks. Chapter 4 examines various roles of NASA and NASA astronauts and provides examples of regulations or policies that apply to similar activities in terrestrial environments, extracting common factors that appear to influence decisions about risk acceptance. Chapter 5 recommends specific ethics principles that are relevant to decisions about human spaceflight. Finally, the report concludes in Chapter 6 by proposing responsibilities applicable to implementing the ethics principles and a decision framework to guide determinations about possible limits and exceptions to health standards for long duration and exploration human spaceflights.

REFERENCES

Angelo, J.A. 2009. *Encyclopedia of space and astronomy*. New York: Infobase Publishing.

Bolden, C. 2013. Presentation at the first meeting of the IOM Committee on Ethics Principles and Guidelines for Health Standards for Long Duration and Exploration Spaceflights. Washington, DC, May 30.

Clarke, A. C. 1968. *The promise of space*. New York: Harper and Row.

Davis, W. 2011. *Into the silence: The Great War, Mallory, and the conquest of Everest*. New York: Alfred A. Knopf.

Dick, S. 2008. *Remembering the space age*. Washington, DC: NASA. http://www.nss.org/resources/library/spacepolicy/Remembering_the_Space_Age.pdf (accessed January 27, 2014).

Dick, S. 2010. *NASA's first 50 years: Historical perspectives*. http://history.nasa.gov/SP-4704.pdf (accessed January 27, 2014).

Dick, S., and K. Cowing. 2004. *Risk and exploration: Earth, sea and the stars*. Presentation at NASA Administrator's Symposium. Monterey, CA, September 26-29. http://history.nasa.gov/SP-4701/riskandexploration.pdf (accessed January 27, 2014).

Duke, M. B. 2013. *Lunar objectives, benefits, pathways, and strategy.* Presentation at the National Research Council's meeting of the Committee on Human Spaceflight. Washington, DC, October 21. http//sites.nationalacademies.org/DEPS/ASEB/DEPS_069080 (accessed March 17, 2014).

Duke, M. B., W.W. Mendell, and B.B. Roberts. 1985. Strategies for a permanent lunar base. In: *Lunar bases and space activities of the 21st century*, edited by W.W. Mendell. Houston, TX: Lunar and Planetary Institute. Pp. 57-68.

Garber, S., and R. Launius. 2005. *A brief history of NASA*. http://history.nasa.gov/factsheet.htm (accessed December 4, 2013).

Harrison, M. 2013. Scurvy on sea and land: Political economy and natural history, c. 1780–c. 1850. *Journal for Maritime Research* 15(1):7-25.

IOM (Institute of Medicine). 2001. *Safe passage: Astronaut care for exploration missions.* Washington, DC: National Academy Press.

IOM. 2004. *Review of NASA's longitudinal study of astronaut health.* Washington, DC: The National Academies Press.

IOM. 2008. *Review of NASA's Human Research Program evidence books: A letter report.* Washington, DC: The National Academies Press.

IOM. 2013. *Environmental decisions in the face of uncertainty*. Washington, DC: The National Academies Press.

Logsdon, J. M., and R. D. Launius. 2008. *Exploring the unknown: Selected documents in the history of the US civil space program, Volume VII.* NASA SP-2008-4407. Washington, DC: NASA.

NASA (National Aeronautics and Space Administration). 2004. *President Bush announces new vision for space exploration program.* http://history.nasa.gov/Bush%20SEP.htm (accessed December 31, 2013).

NASA. 2007. *NASA space flight human system standard. Volume 1: Crew health.* NASA-STD-3001. https://standards.nasa.gov/documents/detail/3315622 (accessed January 27, 2014).

NASA. 2011a. *NASA risk management handbook.* http://www.hq.nasa.gov/office/codeq/doctree/NHBK_2011_3422.pdf (accessed January 27, 2014).

NASA. 2011b. *NASA space flight human system standard. Volume. 2: Human factors, habitability, and environmental health.* NASA-STD-3001. https://standards.nasa.gov/documents/detail/3315785 (accessed December 4, 2013).

NASA. 2013a. *Human exploration and operations at NASA*. http://www.nasa.gov/directorates/heo/home/index.html#.UsHZEPRDsQE (accessed December 30, 2013).

NASA. 2013b. *Commercial space transportation.* http://www.nasa.gov/exploration/commercial (accessed December 30, 2013).

NASA. 2013c. *International Space Station—space station assembly—international partners and participants.* http://www.nasa.gov/mission_pages/station/structure/elements/partners.html (accessed March 11, 2014).

NASA. 2013d. *NASA's management of the commercial crew program.* http://oig.nasa.gov/audits/reports/FY14/IG-14-001.pdf (accessed December 30, 2013).

NASA. 2014a. *Kennedy Space Center visitor complex: Astronaut memorial.* http://www.kennedyspacecenter.com/the-experience/astronaut-memorial.aspx (accessed January 13, 2014).

NASA. 2014b. *New data about the Moon may help create lunar bases.* http://sservi.nasa.gov/articles/lunar-bases (accessed March 12, 2014).

NASA. 2014c. *Yuri Gagarin: First man in space.* http://www.nasa.gov/mission_pages/shuttle/sts1/gagarin_anniversary.html (accessed March 12, 2014).

National Geographic. 2014. *Education: Atmosphere.* http://education.nationalgeographic.com/education/encyclopedia/atmosphere/?ar_a=1 (accessed February 7, 2014).

Norwood, T., and G. Tahu. 2013. *Proposed Mars 2020 mission: NEPA scoping meetings.* Presentation at the National Enviromental Policy Act of 1969 scoping meetings. Washington, DC, September 11.

NRC (National Research Council). 1996. *Understanding risk: Informed decisions in a democratic society.* Washington, DC: National Academy Press.

NRC. 2008. *Science and decisions: Advancing risk assessment.* Washington, DC: The National Academies Press.

NRC. 2011. *Preparing for the high frontier: The role and training of NASA astronauts in the post-space shuttle era.* Washington, DC: The National Academies Press.

NRC. 2013a. *About the Space Studies Board.* http://sites.nationalacademies.org/SSB/ssb_052196 (accessed January 14, 2014).

NRC. 2013b. *Committee on Human Spaceflight statement of task.* http://sites.nationalacademies.org/DEPS/ASEB/DEPS_069080#Statement_of_Task (accessed January 14, 2014).

Solomon, S. 2001. *The coldest march: Scott's fatal Antarctic expedition.* New Haven, CT: Yale University Press.

Stuster, J. 1996. *Bold endeavors: Lessons from polar and space exploration.* Annapolis, MD: Naval Institute Press.

2

NASA Risk Management and Health Standards

Human spaceflight inherently involves a high degree of known risks, as well as uncertain and unforeseeable risks. Risks to astronauts exist during all phases of any space mission, including terrestrial training and vehicle testing, launch, inflight during the mission, and landing. In fact, the launch of the spacecraft is known to be one of the riskiest times in any mission. Health risks during long duration and exploration spaceflights include short-term health consequences (e.g., nausea or fatigue from acute radiation exposure during a solar storm, injury, blurred vision) as well as long-term health consequences that arise or continue months or years after a flight (e.g., radiation-induced cancers; loss of bone mass) (see Chapter 3). The National Aeronautics and Space Administration's (NASA's) Human Research Program has identified 32 space-related health risks that are being studied for possible prevention, treatment, and mitigation approaches (NASA, 2013b). Health standards have been developed by NASA for a number of these risks to set limits on hazardous exposures, outline acceptable health parameters, and guide efforts to protect the health of crew members. This chapter begins with an overview of NASA's approach to risk management and then details the health standards and the process used to develop those standards.

NASA RISK MANAGEMENT PROCESSES

NASA addresses health risks associated with spaceflight using a number of strategies, including engineering, design, mission planning, basic and clinical research, surveillance, medical monitoring, preventive and treatment countermeasures, and health standards. NASA has an extensive research portfolio—managed and implemented through the NASA Human Research Program, National Space Biomedical Research Institute, and other NASA directorates—designed to examine, prevent, and mitigate health and safety risks. The committee was not asked to review NASA's risk management processes but rather to articulate ethics principles, decision points, and recommendations (see Chapters 5 and 6) that should guide health standard decision making surrounding long duration and exploration spaceflight missions and that could be integrated into risk management processes.

Risk management policy at NASA comprises two integrated efforts: risk-informed decision making and continuous risk management processes (NASA, 2008a; Dezfuli, 2013). These processes are used across the agency and apply to engineering, safety, and health standards. Risks are identified based on historical precedence (lessons learned and empirical data), on possible failures in laboratory tests, and in discussions with subject matter experts. Risks are extensively documented and, if possible, quantified. When risk is considered to be unacceptably high, alternative designs and missions scenarios are considered, and the risk assessment continues iteratively. This process is not unique to NASA and represents best practice across a number of industries (INSAG, 2011; FERC, 2013).

In new or emerging areas of concern about space-related health risks, risk assessment may first be based on expert opinion and then be informed by experience gained in a new environment or through case reports and lessons learned from prior space missions (e.g., vision impairment has recently been identified as a risk of spaceflight; see Chapter 3). NASA risk assessment processes initially use a wide margin of uncertainty that, with continued experience, is generally narrowed (Dezfuli, 2013). Decisions regarding the design and implementation of a mission are made based on a combination of available evidence and best risk estimates.

As a federal agency tasked with highly risky missions, NASA must make numerous decisions that balance health and safety risks, technological feasibility, and financial costs against mission necessity and the lost opportunity that comes when missions are not undertaken. NASA states, in part, "an adequately safe system is not necessarily one that completely precludes all conditions that can lead to undesirable consequences" (NASA, 2011c, p. 4). According to NASA policy, adequately safe systems follow two primary safety principles: (1) they meet a minimum threshold level of safety, "as determined by analysis, operating experience, or a combination of both" and aim to improve over time, and (2) they are as "as Safe as Reasonably Practicable" (ASARP) (NASA, 2011c, p. 4). An assessment of whether a system is ASARP involves weighing its safety performance against the impact of the changes that would need to be done to further improve it. "The system is ASARP if an incremental improvement in safety would require a disproportionate deterioration of system performance in other areas" (NASA, 2011c, p. 5). Moreover, informed decision making by those affected is an essential component (Dezfuli, 2013).

Similar to risk assessment for hardware and software, there is a risk assessment process for human health risks, and concomitant strategies for mitigating the risks. By necessity, these two processes are connected and iterative. For example, if radiation exposure in a particular mission is considered too high, this drives the design of the vehicle and the mission design. If there is no engineering or mission design solution to mitigate the risk, then other alternatives are considered, such as redesign of the mission, delays in the mission until technology is available, or making exceptions to the standards. The Bioastronautics Roadmap was developed by NASA as a framework for identifying and assessing the risks to spacecraft crews and included both health and medical risks, as well as engineering technology and system performance risks (NASA, 2005). In the roadmap, "risk" is defined as "the conditional probability of an adverse event from exposure to the space flight environment" and a "risk factor" as a "predisposing condition that contributes to an adverse outcome" (NASA, 2005, p. 3). Specific issues regarding risks were detailed in the roadmap as they pertained to three design reference missions: a 1-year tour of duty on the International Space Station (ISS), a month-long stay on the lunar surface, and a 30-month mission to Mars (NASA, 2005). Design reference missions describe the orbit, mission duration, environments, and proposed operations. There may be multiple design reference missions to a single destination, and each is assessed for challenges, benefits, and risks. These scenarios provide context for mission design and risk identification and assessment, and include information on potential transit times, communication lag times, microgravity-exposure and radiation-exposure levels, vehicle requirements, and the extent of proposed extravehicular activity. The design reference missions continue to be updated and several options for missions to a near-Earth asteroid have been proposed recently along with updates to the Mars design reference mission scenarios (see Table 2-1; NASA, 2009b, 2013c,e).

TABLE 2-1 Examples of NASA's Design Reference Missions

Possible Mission	Approximate Duration of Mission	Surface Stay	Crew Size
Mars	One proposed design is for approximately a 3-year mission (NASA, 2013c)	Short-stay missions: 30 to 90 days (NASA, 2009b); long-stay missions: ~18 months (NASA, 2013c)	6
Near-Earth asteroid (NEA)	One proposed design is for a 12- to 13-month mission with approximately 12 months of transit time and a 1-month stay (NASA, 2013c); other designs depict a 21- or 22-day mission with 16 days of transit, orbit around the NEA of 6 days, and EVAs to the asteroid but no surface stay (NASA, 2013e)	30 days for the year-long mission	3 for the year-long mission
International Space Station	365 days	Not applicable	2[a]
Lunar landing–outpost mission	Approximately 6 months with transit time of 4 days	Approximately 6 months	4

NOTE: EVA = extravehicular activity.

[a]Two crew members would be there for the full year. Other crew members would be there for shorter durations.

SOURCES: NASA, 2009b, 2013c,e.

In 2006, at the request of NASA, the Institute of Medicine (IOM) published a review of the Bioastronautics Roadmap (IOM and NRC, 2006). As part of this comprehensive review, the committee provided several recommendations with respect to format, evidence documentation, definition of specific risks, overarching risk categories (representing potential interaction among risks), and the reframing of risks as either health or technology related. The Bioastronautics Roadmap has since evolved into the Human Research Roadmap, which categorizes 32 health risks into five categories: (1) behavioral health and performance; (2) human health countermeasures (which include bone metabolism and physiology, nutrition, immunology, cardiac and pulmonary physiology, and injury); (3) space radiation; (4) space human factors and habitability; and (5) exploration medical capabilities (NASA, 2013b; see Table 2-2). For each of the 32 areas, the Roadmap provides research reviews and ratings on risk mitigation and control, as discussed below. Chapter 3 provides brief overviews of several of the 32 risks to illustrate how widely the risks vary in the extent to which there are known preventive or treatment measures and to which vehicle or mission design might provide countermeasures.

Each of the risks has been rated by NASA for four design reference missions. The ratings indicate the state of knowledge on mitigating or controlling the risk to meet the NASA standards for maintaining crew performance and health. The ratings system categorizes the risks as: Controlled (C), Acceptable (A), Unacceptable (U), and Insufficient Data (I) (see Table 2-2) (NASA, 2013c). The state of research on each of the 32 risks is summarized in a set of evidence reports, with each report updated and reviewed on a regular basis (NASA, 2013b). These research reviews and ratings provide starting points for identifying gaps and research directions for NASA's Human Research Program, and progress on this research is monitored by NASA's Human System Risk Board. The IOM reviewed NASA's processes for compiling and updating the evidence reports in 2008 (IOM, 2008), and the IOM is conducting a study that will review each evidence report (IOM, 2014).

A number of health risks are known and are mitigated, to the extent feasible, through prevention and intervention strategies, which are continuously evaluated for effectiveness. The seriousness of other likely risks is largely unknown. Still other risks have not been anticipated yet, and will only be uncovered by experience. Thus, setting health and performance standards for this wide array of risks is challenging.

TABLE 2-2 Human Health and Performance Risks with Research Ratings for Design Reference Missions

	HRP Research Rating			
Risk Title	ISS-12	Lunar	NEA	Mars
Risk of orthostatic intolerance during re-exposure to gravity	C	C	C	A
Risk of early onset osteoporosis due to spaceflight	A	A	A	A
Risk factor of inadequate nutrition	C	C	A	U
Risk of compromised EVA performance and crew health due to inadequate EVA suit systems	A	A	A	A
Risk of inadequate performance due to reduced muscle mass, strength and endurance	C	A	A	U
Risk of renal stone formation	C	C	C	C
Risk of bone fracture	C	A	C	A
Risk of intervertebral disc damage	I	I	I	I
Risk of cardiac rhythm problems	A	C	I	I
Risk of reduced physical performance capabilities due to reduced aerobic capacity	C	C	A	U
Risk of crew adverse health event due to altered immune response	C	A	A	A
Risk of impaired control of spacecraft, associated systems and immediate vehicle egress due to vestibular/sensorimotor alterations associated with spaceflight	C	C	C	A
Risk of clinically relevant unpredicted effects of medication	C	C	C	U
Risk of spaceflight-induced intracranial hypertension/vision alterations	U	I	U	U
Risk of decompression sickness	C	C	C	C
Risk of injury from dynamic loads	C	A	A	I
Risk of performance decrement and crew illness due to an inadequate food system	C	C	A	U
Risk of inadequate human-computer interaction	C	C	C	C

	HRP Research Rating			
Risk Title	ISS-12	Lunar	NEA	Mars
Risk of performance errors due to training deficiencies	C	C	C	A
Risk of inadequate design of human and automation/robotic integration	C	C	C	A
Risk of inadequate critical task design	C	C	C	C
Risk of adverse health effects of exposure to dust and volatiles during exploration of celestial bodies	N/A	A	I	I
Risk of incompatible vehicle/habitat design	C	C	C	A
Risk of adverse health effects due to alterations in host-microorganism interactions	C	C	A	A
Risk of unacceptable health and mission outcomes due to limitations of inflight medical capabilities	C	A	A	U
Risk of adverse behavioral conditions and psychiatric disorders	C	C	A	U
Risk of performance errors due to fatigue resulting from sleep loss, circadian desynchronization, extended wakefulness, and work overload	C	C	C	C
Risk of performance decrements due to inadequate cooperation, coordination, communication, and psychosocial adaptation within a team	A	C	A	A
Risk of radiation carcinogenesis	C	A	U	U
Risk of acute radiation syndromes due to solar particle events	C	C	C	C
Risk of acute or late central nervous system effects from radiation exposure	A	A	I	I
Risk of degenerative tissue or other health effects from radiation exposure	A	A	I	I

NOTES: A = Acceptable; C = Controlled; EVA = extravehicular activity; HRP = Human Research Program; I = Insufficient Data; ISS = International Space Station; N/A = not applicable; NEA = near-Earth asteroid; U = Unacceptable.

The research rating categories (full definitions in NASA, 2013c) focus on the extent to which there is evidence that the projected plans for that design reference mission will meet existing standards for maintaining crew health and performance or that countermeasures exist to control the risk.

SOURCE: NASA, 2013c.

HEALTH STANDARDS

The use of health standards to protect individuals engaged in specific types of work or activities is not unique to space exploration or exploratory activities. A standard is a requirement, norm, or convention that applies to performing an activity or to the properties of a product or thing. Governments have long applied health standards to protect the welfare of workers (see Chapter 4). In occupational settings, health standards guide design, research, and engineering activities and commonly serve the following core functions:

- **Worker Protection:** Health standards, often in the form of permissible exposure limits or performance standards, limit the duration and/or level of exposure to hazardous substances or set the health or performance levels that workers need to maintain. For example, OSHA issues permissible exposure limits that set regulatory limits on hazardous exposures for a wide range of workplaces and industries (OSHA, 2013).
- **Stimulation of Design and Innovation:** Health standards establish the specifications toward which engineers, researchers, and others strive in their design and innovation. In the hierarchy of worker protection controls against occupational hazards, the preferred solution is to eliminate the hazard. However, if that is not possible, then an engineering, design, or other solution is sought that can be built into the work environment and, therefore, does not require individuals behaviors. For example, individuals would need to remember to arrange work schedules to limit the time a worker is exposed (an example of an administrative control) or to put on gloves or respiratory protection (examples of personal protective equipment controls). Standards provide the goals for engineering or operational solutions during the design and development of the mission. Standards also provide guidance on flight rules for implementation during the mission.
- **Level the Playing Field:** Health standards can also be used as transparent criteria for specific jobs, so that prospective workers and employers have a clear understanding of the job requirements. In this respect, standards are instrumental in ensuring that everyone knows from the outset the criteria for the position and all meet the requirements. Criteria for applying to the astronaut corps are discussed later in this chapter.

- **Engendering Confidence in, and Legitimacy for, Institutions:** Standards also play a role in stabilizing collaborative human activities. For instance, space exploration requires large outlays of material and infrastructural support, much of which derives from federal funding. As astronauts are not conscripted, it also requires the voluntarism of individuals who join the astronaut corps and agree to participate in spaceflight missions. Stewardship of this support depends in part on ensuring that the institutions are acting prudently, competently, and showing due regard for those who enter into the collaboration.

Health Standards for Spaceflight

In its efforts to protect the health and safety of spaceflight crew members, NASA has developed spaceflight human system standards (termed in this report as "health standards"), as delineated in Boxes 2-1 and 2-2, that aim to provide a "healthy and safe environment for crew members, and to provide health and medical programs for crew members during all phases of space flight" (NASA, 2007, p. 8). These standards include consideration of preflight, inflight, and postflight health issues and comprise two volumes: *Volume 1: Crew Health* and *Volume 2: Human Factors, Habitability, and Environmental Health* (NASA, 2007, 2011b) (see Boxes 2-1 and 2-2).[1] NASA's health standards fall into three categories:

(1) Fitness-for-duty standards—These standards provide a "minimum measurable capability or capacity" for the specific parameter (e.g., aerobic capacity) based in some cases on normative values for age and sex for the general population (NASA, 2007, p. 17).

(2) Space permissible exposure limits—These types of standards set ceilings on risk exposures during missions and are based on measures of exposure to a physical or chemical agent (e.g., radiation) with quantifiable limits over a given amount of time (e.g., lifetime radiation exposure).

---

[1]NASA's health standards are complemented by NASA standards for engineering and for safety and mission assurance with documentation, including the *Human Integration Design Handbook*, *NASA Spacecraft Maximum Allowable Concentration Tables*, and *NASA Spacecraft Water Exposure Guidelines* (NASA, 2008b,c, 2010, 2012b).

(3) Permissible outcome limits—These standards reference the assessment of a biological or clinical parameter with the standard outlining the acceptable maximum decrease or change (e.g., bone density) (NASA, 2007).

The standards may be cited in NASA contract, program, and other documents as technical requirements; "mandatory requirements are indicated by the word 'shall,' statement of fact and descriptive material by 'is,' and permission by 'may' or 'can.' Tailoring of, deviation from, or waivers to this standard for application to a specific program or project shall be approved by the NASA Chief Health and Medical Officer" (NASA, 2007, p. 9).

NASA develops the standards to cover the broad range of space missions. The agency also develops more detailed documents for each spaceflight, including the Crew Health Concept of Operations document and the Medical Operations Requirements document (NASA, 2007).

BOX 2-1
NASA Health Standards: Crew Health (2007)

1. **Fitness-for-Duty Aerobic Capacity Standard**
   a. Crew members shall have a pre-flight maximum aerobic capacity (VO_{2max}) at or above the mean for their age and sex (see American College of Sports Medicine Guidelines below in Table A).
   b. The in-flight aerobic fitness shall be maintained, either through countermeasures or work performance, at or above 75 percent of the pre-flight value, as determined by either direct or indirect measures.
   c. The post-flight rehabilitation shall be aimed at achieving a VO_{2max} at or above the mean for age and sex (see Table A).

TABLE A 50th Percentile Values for Maximal Aerobic Power (ml kg^{-1} min^{-1})

Age	Men	Women
20-29	43.5	35.2
30-39	41.0	33.8
40-49	38.1	30.9
50-59	35.2	28.2
60+	31.8	25.8

2. **Fitness-for-Duty Sensorimotor Standard**
   a. Pre-flight sensorimotor functioning shall be within normal values for age and sex of the astronaut population.
   b. In-flight fitness-for-duty standards shall be guided by the nature of mission-associated high-risk activities, and shall be assessed using metrics that are task specific.

c. Sensorimotor performance limits for each metric shall be operationally defined.
d. Countermeasures shall maintain function within performance limits.
e. Post-flight rehabilitation shall be aimed at returning to baseline sensorimotor function.

3. **Fitness-for-Duty Behavioral Health and Cognition Standard**
   a. Pre-flight, in-flight, and post-flight crew behavioral health and crew member cognitive state shall be within clinically accepted values as judged by clinical psychological evaluation.
   b. End-of-mission rehabilitation for crew member cognitive state shall be aimed at transitioning the crew member back to pre-flight values.
   c. End-of-mission rehabilitation for behavioral health of the crew member shall be aimed at transitioning the crew member back into terrestrial work, family, and society.
   d. The planned number of hours for completion of critical tasks and events, workday, and planned sleep period shall have established limits to assure continued crew health and safety.

4. **Fitness-for-Duty Hematology and Immunology Standard**
   a. Pre-launch hematological/immunological function shall be within normative ranges established for the healthy general population.
   b. In-flight countermeasures shall be in place to sustain hematological/immunological parameters within the normal range as determined by direct or indirect means.
   c. Countermeasures and monitoring shall be developed to ensure immune and hematology values remain outside the "critical values" (i.e., that level which represents a significant failure of the hematological/immunological system and is associated with specific clinical morbidity) defined for specific parameters.
   d. Post-flight rehabilitation shall be aimed at returning to pre-flight baseline.

5. **Permissible Outcome Limit for Nutrition Standard**
   a. Pre-flight nutritional status shall be assessed and any deficiencies mitigated prior to launch.
   b. In-flight nutrient intake shall be no less than 90 percent of the calculated nutrient requirements, based on an individual's age, sex, body mass (kg), height (m), and an activity factor of 1.25.
   c. Nutrient planning shall be aimed at maintaining a body mass and composition greater than 90 percent of pre-flight values.
   d. Post-flight nutritional assessment and rehabilitation shall be aimed at returning to baseline.

6. **Permissible Outcome Limit for Muscle Strength Standard**
   a. Pre-flight muscle strength and function shall be within normal values for age and sex of the astronaut population.
   b. Countermeasures shall maintain in-flight skeletal muscle strength at or above 80 percent of baseline values.
   c. Post-flight rehabilitation shall be aimed at returning to baseline muscle strength.

7. **Permissible Outcome Limit for Microgravity-Induced Bone Mineral Loss Performance Standard (Baseline with Measured T-Score)**
   a. Crew members' pre-flight bone mass Dual Energy X-ray Absorptiometry (DEXA T) score shall not exceed -1.0 (-1.0 Standard Deviation [SD] below the mean Bone Mineral Density).
   b. Countermeasures shall be aimed at maintaining bone mass in-flight consistent with outcome limits.
   c. The post-flight (end of mission) bone mass DEXA T score shall not exceed -2.0 (-2.0 SD below the mean Bone Mineral Density).
   d. Post-flight rehabilitation shall be aimed at returning bone mass to pre-flight baseline.

8. **Space Permissible Exposure Limit for Space Flight Radiation Exposure Standard**
   a. Planned career exposure for radiation shall not exceed 3 percent risk of exposure-induced death (REID) for fatal cancer.
   b. NASA shall assure that this risk limit is not exceeded at a 95 percent confidence level using a statistical assessment of the uncertainties in the risk projection calculations to limit the cumulative effective dose (in units of Sievert) received by an astronaut throughout his or her career.
   c. Exploration Class Mission radiation exposure limits shall be defined by NASA based on National Council on Radiation Protection (NCRP) recommendations.
   d. Planned radiation dose shall not exceed short-term limits as defined in Table B.
   e. In-flight radiation exposures shall be maintained using the as low as reasonably achievable (ALARA) principle.

TABLE B from Appendix F Dose limits for short-term or career non-cancer effects (in mGy-Eq or mGy). Note RBE's for specific risks are distinct as described below.

Organ	30-day limit	1-year limit	Career
Lens*	1,000 mGy-Eq	2,000 mGy-Eq	4,000 mGy-Eq
Skin	1,500	3,000	4,000
BFO	250	500	Not applicable
Heart**	250	500	1,000
CNS***	500	1,000	1,500
CNS*** ($Z \geq 10$)	—	100 mGy	250 mGy

NOTES: BFO = blood-forming organ; CNS = central nervous system; RBE = relative biological effectiveness. Supporting information and rationale for each of the standards is found in Appendix F, Rationale for Space Flight Health Standards for Human Performance, of NASA-STD-3001 (NASA, 2007).

*Lens limits are intended to prevent early (<5 yr) severe cataracts (e.g., from a solar particle event). An additional cataract risk exists at lower doses from cosmic rays for sub-clinical cataracts, which may progress to severe types after long latency (>5 yr) and are not preventable by existing mitigation measures; however, they are deemed an acceptable risk to the program.

**Heart doses calculated as average over heart muscle and adjacent arteries.

***CNS limits should be calculated at the hippocampus.

SOURCE: NASA, 2007.

BOX 2-2
Examples of NASA Health Standards on Human Factors, Habitability, and Environmental Health (2011)

- **Physical Characteristics and Capabilities**
  - **Aerobic Capacity (4.9):** An individual's absolute aerobic capacity determines the ability to perform a task at a given level of work. The system shall be designed to be operable by crew members with the aerobic capacity as defined in NASA-STD-3001, Volume 1.
- **Perception and Cognition**
  - **Situational Awareness (SA) (5.2.2):** SA refers to the process and outcome of understanding the current context and environment, evaluating that situation with respect to current goals, and projecting how that situation will evolve in the future. Systems shall be designed such that the SA necessary for efficient and effective task performance is provided and can be maintained for all levels of crew capability and all levels of task demands.
- **Natural and Induced Environments**
  - **Atmospheric Data Recording (6.2.6.1):** For each isolatable, habitable compartment, the system shall automatically record pressure, humidity, temperature, ppO_2, and $ppCO_2$ data.
- **Habitability Functions**
  - **Orthostatic Intolerance Countermeasures (7.4.5):** The system shall provide countermeasures to mitigate the effects of orthostatic intolerance when transitioning from microgravity to gravity environments.
- **Architecture**
  - **Consistent Orientation (8.2.2.2):** In microgravity, the system shall establish a local vertical orientation.
- **Crew Interfaces**
  - **Private Audio Communication (10.5.3.9):** The system shall provide the capability for private audio communication with the ground.
- **Spacesuits**
  - **Suit Equilibrium Pressure (11.1.3.2):** Suits shall maintain pressure within 0.1 psi (0.689 kPa) after the suit has achieved an equilibrium pressure for a set-point.

SOURCE: NASA, 2011b.

Developing and Updating NASA's Health Standards

NASA's health standards are established and maintained by NASA's Office of the Chief Health and Medical Officer (OCHMO) who reports to the NASA Administrator (NASA, 2013d). The mandate for these standards is delineated in NASA Policy Directive 8900.5, which specifies that NASA's policy is to "provide a healthy and safe environment for crewmembers to enable successful human space exploration" with the OCHMO responsible for establishing and maintaining spaceflight health and medical standards (NASA, 2011a, p. 1). The standards follow an occupational health model that sets hazardous exposure limits and delineates health criteria for workers.

Requests for new or revised health standards are submitted to the OCHMO which decides whether or not a standards development team will be assembled. Any office within NASA, including the Astronaut Office,[2] may initiate a request. Standards development teams include internal NASA experts, and external experts can be appointed (NASA, 2012a). If the OCHMO decides that the revision of an existing standard or the development of a new standard should be explored, the standards development team first conducts a comprehensive review of the available scientific and clinical evidence, as well as operational data and experience from Apollo, Skylab, Shuttle, Shuttle-Mir, and ISS missions. The team then prepares a draft (or revised) standard that is reviewed by the Chief Health and Medical Officer who decides whether to convene an external technical review (NASA, 2007, 2012a). After these reviews are completed, the revised or new standard is presented to the NASA Medical Policy Board, which decides whether to recommend the standard to the Chief Health and Medical Officer, who is responsible for final approval (NASA, 2012a; Liskowsky, 2013). A 2007 IOM report commissioned by NASA assessed the health standards development process and outlined a set of principles for standard-setting, specifying that they should be evidence-based, open and transparent, well documented, well informed, and dynamic (IOM, 2007).

As with all NASA standards, reviews of the health standards are conducted every 5 years (NASA, 2012a). Reviews of the health stand-

---

[2]Comprised of the members of the Astronaut Corps (astronauts selected and trained to fly as crew members) as well as administrative and engineering support staff, the Astronaut Office provides support for safety reviews, engineering tests, program development, public and educational outreach, and government and external collaborations (NRC, 2011). The Chief of the Astronaut Office reports to the Director of Flight Crew Operations.

ards can also be conducted at any time that new research data or information from clinical observations indicate that an update review is needed (NASA, 2012a). Inputs into NASA's review decision often include studies from external organizations, including the National Research Council Committee on Toxicology's review of exposure guidelines for selected airborne contaminants and the IOM's reviews of the Human Research Program's evidence reports (e.g., NRC, 2008; IOM, 2014). NASA's Human System Risk Board also monitors the status of each of the health risks.

NASA's health standards apply to all NASA human spaceflight programs including long duration and exploration missions (NASA, 2007). NASA standards "address those areas where the human system has shown particular vulnerability in response to adaptation or exposure to microgravity" (NASA, 2007, p. 10). The ethics challenges that result from having one set of space health standards across a wide swath of mission types are considered later in this report (see Chapters 5 and 6). As noted in the 2007 IOM report, "The challenge in developing space flight health standards is to determine an acceptable level of risk that provides maximum feasible protection of crew health and safety without jeopardizing mission success or 'overengineering' either the technical or medical solutions that mitigate these risks. This challenge becomes more daunting as missions become more complex, involving new vehicles, longer durations, greater distances from Earth, and novel environments" (IOM, 2007, p. 5). The report noted the paucity of data for many of the health risks and highlighted the challenges posed by the need to conduct quantitative risk assessments, limited data regarding some health issues or hazardous exposures, and the small number of individuals who have experienced spaceflight, particularly regarding certain exposures (IOM, 2007).

NASA Crew Selection and Medical Certification Standards

In addition to spaceflight health standards, NASA has developed standards that specify health-related criteria for astronaut selection and parameters for health and medical screening, evaluation, and certification including annual certification medical examinations (NASA, 2012a). Applicants to NASA's Astronaut Candidate Program, in addition to meeting specified academic and citizenship requirements, must complete long duration spaceflight physicals with requirements to meet a number of criteria, including distance and near visual acuity correctable to 20/20 in each eye (refractive surgical procedures of the eye are allowed, "providing at least 1 year has passed since the date of the procedure with no permanent adverse after effects"), blood pressure below 140/90 in a sitting position, and specific height requirements (NASA, 2013a).

Once in the program, astronauts have annual medical examinations and screenings. As needed, waivers on the medical criteria for annual certification are considered by the NASA Aerospace Medicine Board on a case-by-case basis to determine if an astronaut who develops a health condition can continue to work (NASA, 2009a). Waivers are used to address issues specific to an individual astronaut's risk profile but are not designed for situations where the mission risk fails to meet accepted parameters.

SUMMARY AND RECOMMENDATION

NASA has detailed risk management processes applicable to its efforts in engineering, safety, and health. The starting point in decisions about whether to allow levels of risks that would be in excess of NASA's health standards is determining that the current health standards reflect the most relevant and up-to-date evidence about spaceflight-related risks to human health and safety. The committee believes it is important for NASA to articulate the criteria for evaluating new evidence and explicitly outline the processes it uses to ensure that evaluation, implementation, and potential revision of health standards are fully consistent with the set of ethics principles outlined in this report. Current NASA policy does specify the administrative processes and levels of approval for the initiation of a new health standard or a revision to a current health standard. Health standards are reviewed every 5 years or reviews can be triggered at any time if new research data or information from clinical observations indicate an update review is needed. However, the committee believes additional information about decision criteria, process, and application of ethics principles should be available to the public.

> **Recommendation 1: *Expand on the Policies for Initiating and Revising Health Standards***
> **NASA should ensure that its policies regarding health standards detail the conditions or circumstances (and relevant priorities) that initiate development or revision of health standards and explicitly indicate how these policies are fully consistent with the set of ethics principles outlined in this report.**

REFERENCES

Dezfuli, H. 2013. *Evolution of risk management at NASA and the philosophy of risk acceptance*. PowerPoint presented at the second meeting of the Institute of Medicine Committee on Ethics Principles and Guidelines for Health Standards for Long Duration and Exploration Spaceflights. Washington, DC, July 25. http://www.iom.edu/~/media/Files/Activity%20Files/Research/HealthStandardsSpaceflight/2013-JUL-25/Panel%201%20-%20Dezfuli.pdf (accessed October 18, 2013).

FERC (Federal Energy Regulatory Commission). 2013. Risk-informed decision-making. http://www.ferc.gov/industries/hydropower/safety/initiatives/risk-informed-decision-making/about.asp (accessed February 19, 2014).

INSAG (International Nuclear Safety Group). 2011. A *framework for an integrated risk informed decision making process*. INSAG-25. http://www.pub.iaea.org/MTCD/publications/PDF/Pub1499_web.pdf (accessed February 19, 2014).

IOM (Institute of Medicine). 2007. *Review of NASA's space flight health standards-setting process: Letter report*. Washington, DC: The National Academies Press.

IOM. 2008. *Review of NASA's Human Research Program evidence books: A letter report*. Washington, DC: The National Academies Press.

IOM. 2014. *Review of NASA's evidence reports on human health risks: 2013 letter report*. Washington, DC: The National Academies Press.

IOM (Institute of Medicine) and NRC (National Research Council). 2006. *A risk reduction strategy for human exploration of space: A review of NASA's Bioastronautics Roadmap*. Washington, DC: The National Academies Press.

Liskowsky, D. R. 2013. *NASA health standards*. PowerPoint presented at the first meeting of the Institute of Medicine Committee on Ethics Principles and Guidelines for Health Standards for Long Duration and Exploration Spaceflights. Washington, DC, May 30. http://www.iom.edu/~/media/Files/Activity%20Files/Research/HealthStandardsSpaceflight/David%20Liskowsky.pdf (accessed November 8, 2013).

NASA (National Aeronautics and Space Administration). 2005. *Bioastronautics roadmap: A risk reduction strategy for human space exploration*. NASA/SP-2004-6113. http://humanresearchroadmap.nasa.gov/into/BioastroRoadmap.pdf (accessed December 4, 2013).

NASA. 2007. *NASA space flight human system standard. Volume 1: Crew health*. NASA-STD-3001. https://standards.nasa.gov/documents/detail/3315622 (accessed December 4, 2013).

NASA. 2008a. *Agency risk management procedural requirements*. NPR 8000.4A. http://nodis3.gsfc.nasa.gov/npg_img/N_PR_8000_004A_/N_PR_8000_004A_.pdf (accessed December 4, 2013).

NASA. 2008b. *Spacecraft maximum allowable concentrations for airborne contaminants*. JSC-20584. http://www.nasa.gov/centers/johnson/pdf/485930main_SMACsGuidelines.pdf (accessed December 6, 2013).

NASA. 2008c. *Spacecraft water exposure guidelines (SWEGs)*. JSC-63414. http://www.nasa.gov/centers/johnson/pdf/485931main_SWEGsGuidelines.pdf (accessed December 6, 2013).

NASA. 2009a. *NASA aviation medical certification standards*. OCHMO 110902.2MED. http://www.nasa.gov/pdf/620882main_Av_med_cert_std_OCHMO_110902_2MED.pdf (accessed December 6, 2013).

NASA. 2009b. *Human exploration of Mars: Design reference architecture 5.0*. NASA/SP-2009-566. http://www.nasa.gov/pdf/373665main_NASA-SP-2009-566.pdf (accessed January 8, 2014).

NASA. 2010. *Human integration design handbook (HIDH)*. NASA/SP-2010-3407. http://ston.jsc.nasa.gov/collections/trs/_techrep/SP-2010-3407.pdf (accessed December 6, 2013).

NASA. 2011a. *NASA health and medical policy for human space exploration*. NPD 8900.5B. http://nodis3.gsfc.nasa.gov/npg_img/N_PD_8900_005B_/N_PD_8900_005B__main.pdf (accessed December 4, 2013).

NASA. 2011b. *NASA space flight human system standard. Volume 2: Human factors, habitability, and environmental health*. NASA-STD-3001. https://standards.nasa.gov/documents/detail/3315785 (accessed December 4, 2013).

NASA. 2011c. *NASA system safety handbook: Volume 1, System safety framework and concepts for implementation*. SP-2010-580. http://www.hq.nasa.gov/office/codeq/doctree/NASASP2010580.pdf (accessed October 22, 2013).

NASA. 2012a. *NASA health and medical requirements for human space exploration*. NASA Procedural Requirements 8900.1A. http://nodis3.gsfc.nasa.gov/npg_img/N_PR_8900_001A_/N_PR_8900_001A_.pdf (accessed December 5, 2013).

NASA. 2012b. *Standards and technical assistance resource tool (START)*. https://standards.nasa.gov (accessed December 6, 2013).

NASA. 2013a. *Astronaut candidate program*. http://astronauts.nasa.gov/content/broch00.htm (accessed December 6, 2013).

NASA. 2013b. *Human research program evidence*. http://humanresearchroadmap.nasa.gov/evidence (accessed October 22, 2013).

NASA. 2013c. *Human research program requirements document, Revision F*. HRP-47052. http://www.nasa.gov/pdf/579466main_Human_Research_Program_Requirements_DocumentRevF.pdf (accessed October 18, 2013).

NASA. 2013d. *NASA organizational chart*. http://www.nasa.gov/about/org_index.html#.Up-YeMRDt8F (accessed December 4, 2013).

NASA. 2013e. *Asteroid redirect mission and human space flight*. Presentation to the National Research Council Committee for Study on Human Space Flight Technical Panel, June 19. http://www.nasa.gov/pdf/756678main_20130619-NRC_Tech_Panel_Stich.pdf (accessed January 13, 2014).

NRC (National Research Council). 2008. *Spacecraft maximum allowable concentrations for selected airborne contaminants*. Washington, DC: The National Academies Press.

NRC. 2011. *Preparing for the high frontier: The role and training of NASA astronauts in the post-space shuttle era*. Washington, DC: The National Academies Press.

OSHA (Occupational Safety and Health Administration). 2013. *Permissible exposure limits*. https://www.osha.gov/dsg/topics/pel (accessed November 1, 2013).

3

Health Risks

In considering the ethics issues that will emerge when making decisions about sending humans into harm's way on long duration and exploration spaceflights, the committee decided to examine some of the health risks that illustrate key ethical challenges and tensions in risk and decision making. It is important to note that the stressors and health risks may vary from mission to mission, depending on the remoteness of the destination and many other factors. For example, missions to locations that will require months to years to return to Earth can have fundamentally different stressors than destinations that are closer but may involve being in space for long periods of time. Although the phrase "long duration and exploration spaceflights" is used throughout the report, the committee acknowledges that the health risks may vary widely between missions.

These examples are meant to illustrate the diversity of risks and, therefore, highlight some of the ethics quandaries arising in decisions regarding uncertain risks, unknowns about the extent of individual variations, and long-term health implications for astronauts. In addition, the examples also demonstrate the range of risk management strategies that are being employed by or considered by the National Aeronautics and Space Administration (NASA), international partners, and commercial companies to prevent, mitigate, or treat these risks to human health. The following sections are brief overviews of the complex issues relevant to:

- the risk of loss of life due to mission operations,
- vision impairment risks,
- behavioral health and performance risks,

- bone demineralization, and
- radiation exposure.

MISSION OPERATIONS RISKS

Although there have been many successful space missions over the past 50 years, the substantial risks undertaken by astronauts are most evident in documented "near misses" and tragic losses of life. During the Apollo 11 mission, Neil Armstrong piloted a lunar module to the Moon's surface, landing with less than 30 seconds worth of fuel remaining (Garber and Launius, 2005). Apollo 12 was struck by lightning at approximately 36 and 52 seconds after launch, momentarily shutting down the electrical power (Molloy and Petrone, 2013). During the Apollo 13 mission, an oxygen tank exploded while en route to the Moon, and ground crews and astronauts had to improvise to safely return the crew to Earth (Garber and Launius, 2005). The 1997 collision of a resupply vehicle, as well as a fire, jeopardized the lives of crew members aboard the Mir Space Station (NASA, 2014b). In 1967, a cabin fire in an Apollo capsule during a launch pad test in Cape Canaveral, Florida, killed three astronauts.[1] Over the course of the NASA space program, 24 crew members have lost their lives in the line of duty, including the 14 individuals aboard the space shuttles Challenger and Columbia, the Apollo 1 astronauts, and 5 individuals who died in training-jet crashes (NASA, 2014a). Due to the nature of the propulsion systems needed to launch, the distance from Earth, the immediacy of some inflight catastrophes and the challenges of landing, options for rescue operations are limited.

The probability of loss of crew due to a catastrophic vehicle accident is generally higher and accompanied by greater uncertainty in early experience with new crew delivery vehicles. As experience is gained through effective use of continuous risk management systems, the expected results are decreased risk and less uncertainty, although risks remain high. Thus, for example, the probabilistic risk assessment for loss of crew and vehicle on the first shuttle flight was 1/10 versus 1/90 on the 135th shut-

[1]On January 27, 1967, the crew of Roger B. Chaffee, Virgil "Gus" Grissom, and Edward H, White, Jr., were killed in a fire that spread quickly in a pure oxygen environment; the astronauts had no opportunity to open the hatch (Garber and Launius, 2005; Williams, 2011).

tle flight (Behnken et al., 2013). The shuttle mission aggregate risk of loss of crew and vehicle was approximately 1/46 (Behnken et al., 2013).

The risks of the loss of the crew and vehicle during a spaceflight mission are calculated to allow comparisons of different approaches and platforms using probabilistic risk assessments. NASA provides detailed briefings for astronaut candidates on mission risks. Comparisons are made in this briefing to the risk of death in other high-risk occupations and to catastrophic events and combat. Astronauts who provided information to the committee emphasized the thorough and ongoing nature of communication throughout their careers regarding risks.

VISION IMPAIRMENT

Vision impairment represents a newly identified health risk and exemplifies how NASA approaches a new risk in the current risk management framework. The potential for vision changes has both acute and long-term implications for the astronaut and for current and future missions.

Overview and Risk Identification

Although astronauts have reported vision changes during spaceflight for more than 40 years (Alexander et al., 2012), these were assumed to be transient and isolated. With the Mir Space Station and the International Space Station (ISS), longer tours of duty became possible. A Mir technical report from 2008 described disc edema in 8 of 16 cosmonauts on landing, with one magnetic resonance imaging (MRI) report showing signs suggestive of intracranial hypertension (as cited in Alexander et al., 2012). The spaceflight environment on Mir was noted to be similar to the ISS.

More significant and lasting visual changes have occurred in astronauts on the ISS and have been documented in published case reports (Mader et al., 2011). These changes were primarily a shift toward hyperopia (far-sightedness). Scotomas (blind spots in the visual field) were also reported. One astronaut reported that he needed to shift his head to compensate for scotoma when reading instruments during the mission (Alexander et al., 2012). In a group of astronauts examined pre- and postflight, clinicians observed flattening of the eyeball globe using ultrasound and MRI, supporting the hyperopic shift, edema of the optic nerve, and changes of the optic fundus noted on examination (Mader et al., 2011). In this discussion of the detailed findings on seven astronauts after spaceflight, the results are presented in an "unattributable" manner and do not include dates or missions to preserve privacy and confidentiality (Mader et al., 2011). Refractive changes have persisted after return to Earth for these astronauts.

Upon identifying the vision-related issues, NASA rapidly implemented pre- and postflight ocular testing protocols and convened a Papilledema Summit in July 2009 to bring together experts in space medicine and terrestrial analogs and to suggest avenues for research (Watkins and Barr, 2010). The summit examined potential physiological causes of the vision alterations and suggested changes in preflight, inflight, and postflight testing.

Because the optic changes bore a striking similarity to those seen with elevated intracranial pressure (i.e., papilledema), this hypothesis was the first to be extensively explored (Alexander et al., 2012). Lumbar puncture has been performed on four U.S. astronauts (at single intervals ranging from 12 to 60 days after flight), and opening pressures were found to be elevated in two individuals and borderline in the other two (Alexander et al., 2012). Other hypotheses regarding causes of vision alterations include elevated carbon dioxide in the ISS atmosphere, increased sodium content of the astronaut diet, fluid shifts due to increases in resistance exercises, radiation exposure, and individual susceptibility (Alexander et al., 2012).

Since 1989, as part of postflight eye examinations, U.S. astronauts have been asked about improvements or degradations in their distant or near vision. Compilation of that information on 300 U.S. astronauts who had flown in space found that approximately 29 percent of shuttle astronauts and 60 percent of ISS astronauts noted a degradation in distant and near visual acuity, and some of these changes remained unresolved years after flight (Mader et al., 2011). With the exception of the report from Mir previously cited, data from international partners are lacking.

Risk Management: Clinical Practice Guidelines, Surveillance, and Research

NASA is monitoring vision changes and intracranial pressure, and research efforts in this area are fully under way. NASA has developed clinical practice guidelines for treatment of astronauts with postflight refractive changes, which include a classification system from class 0 (least severe) through class 4 (most severe) based on the results of imaging studies and indicating the follow-up testing and monitoring that are required (Alexander et al., 2012).

In 2012, the NASA Human Research Program released the evidence report *Risk of Spaceflight-Induced Intracranial Hypertension and Vision Alterations*, which summarizes research evidence and raises questions for future study (Alexander et al., 2012). Current research efforts are exploring ground-based analogs and include studies of normal individuals in supine and head-down bed rest, hind-limb-suspended rodent models, and various pathologic situations in which elevated intracranial hypertension and papilledema occur. The evidence report details four major gaps requiring future research:

- Etiological mechanisms and contributing risk factors for ocular structural and functional changes seen inflight and postflight;
- Validated and minimally obtrusive diagnostic tools to measure and monitor changes in intracranial pressure, ocular structure, and ocular function;
- Ground-based analogs and/or models to simulate the spaceflight-associated visual impairment and increased intracranial pressures; and
- Preventive and treatment measures to mitigate changes in ocular structure and function and intracranial pressure during spaceflight (Alexander et al., 2012).

Individual Variation Issues

As with many health risks, individual factors (including, but not limited to, age and sex) may account for some of the variability in manifestation of vision and ocular alterations. The evidence report on this risk concludes with this statement, "In summary, 15 long-duration male astronauts ranging in age from 45 to 55 years have experienced confirmed inflight and postflight visual and anatomical changes" (Alexander et al., 2012, p. 87).

Health Standards and Risk Profile

A NASA health standard has not yet been developed for vision impairment and increased intracranial pressure and the extent, if any, to which these are linked is not fully known (IOM, 2014). At present, this risk is listed as "unacceptable" on the Human Research Roadmap (see Table 2-2) for missions to the ISS for 12 months, to a near-Earth asteroid, or to Mars, and "insufficient data" for a lunar mission (NASA, 2013).

Issues for Long Duration and Exploration Spaceflights

In 2013, Mader and colleagues reported on a single astronaut who was carefully studied after two long-duration missions, documenting progressive vision changes associated with repeat spaceflight, raising concerns that changes occurring during the first flight "may have set the stage for recurrent or additional changes when the astronaut was subjected to physiological stress of repeat space flight" (Mader et al., 2013, p. 249). The prevalence and severity of the observed visual changes raised significant concerns about the ability of an affected pilot to successfully land a spacecraft. This is an active area of research with many remaining unknowns that could affect future crew selection, mission success, and astronaut health.

Ethics issues raised by this example focus on how best to address uncertainty and unknowns including determinations of long-term responsibilities for monitoring, preventing, and treating the health condition moving forward. Issues of individual susceptibility are still being determined, but this example points to the importance of diverse participation, including women, to obtain population-based information (as all examples to date of vision problems have occurred in men).

BEHAVIORAL HEALTH AND PERFORMANCE RISKS

NASA has identified three categories of behavioral health and performance risks associated with long duration and exploration spaceflight: (1) adverse behavioral conditions and psychiatric disorders; (2) performance errors due to fatigue resulting from sleep loss, circadian desynchronization, extended wakefulness, and work overload; and (3) performance decrements due to inadequate cooperation, coordination, communication, and psychosocial adaptation within a Team Gap (Schmidt et al., 2009; Slack et al., 2009; Whitmire et al., 2009).

Overview and Risk Identification

Much of the evidence regarding behavioral health risks associated with long duration spaceflight is derived from anecdotal evidence (Aldrin, 1973; Lebedev, 1988; Burrough, 1998; Linenger, 2000), archival and observational data collected during spaceflight (Kanas et al., 2007; Stuster, 2010), and observational and experimental studies conducted in analog settings, such as polar expeditions (Gunderson, 1974; Sandal et al., 1996; Palinkas and Suedfeld, 2008), submarines (Tansey et al., 1979; Sandal et al., 1999; Thomas et al., 2000), and space simulations (Gushin et al., 1996; Sandal, 2001; Basner et al., 2013). The Mars 500 study, with participants from four nations, was a 17-month isolation experiment in preparation for a mission to Mars (ESA, 2011).

While the reported incidence of behavioral health problems encountered during spaceflight has been quite low, the actual incidence may be underestimated due to a reluctance of astronauts to report them (IOM, 2001; Shepanek, 2005). Billica (2000) reported a 2.86 per person-year incidence of such problems among the 508 crew members who flew on 89 space shuttle missions between 1981 and 1989. The most common behavioral symptoms reported by crew members were anxiety and irritability. Data collected for 28.84 person-years of NASA spaceflight identified 24 cases of anxiety, for an incidence rate of 0.832 cases per person-year (Slack et al., 2009). No behavioral emergencies have been reported to date (Slack et al., 2009).

Studies in analog settings have reported much higher rates of behavioral health problems, reflecting longer periods of isolation and confinement and differences in crew member characteristics. For example, the incidence of behavioral health problems after extended stays in Antarctica was estimated in one study at 5.2 percent (Palinkas et al., 2004).

A number of factors may contribute to the risk of behavioral health and sleep problems in space, including mission duration (Slack et al., 2009), disruption of sleep and circadian rhythms (Czeisler et al., 1986; Dinges et al., 1997), physiological changes that occur in microgravity, workload, lack of social and environmental stimulation, cultural and organizational factors, family issues, and personality characteristics (Gunderson, 1974; McFadden et al., 1994; Rose et al., 1994; NRC, 1998; Rosnet et al., 2000). For the ISS, both observational (i.e., Stuster, 2010) and anecdotal evidence indicate that the crew has been successful at thriving.

Risk Management: Countermeasures and Research

Analog and other types of studies have identified measures that can prevent or mitigate behavioral health, sleep impairment, and cognition risks. Countermeasures that have been successfully implemented for ISS operations include working with ground control regarding scheduling, support services from operational psychology personnel, multiple layers of accountability when conducting critical tasks, and education for ground crews (Flynn, 2005; Slack et al., 2009). Existing techniques for monitoring crew behavioral health and providing social and psychological support also have proved effective on the ISS. However, it is unknown whether the effectiveness of these countermeasures can be maintained over longer periods and at greater distances from Earth where delays in communication, delivery of services, and implementation of countermeasures will occur. Policies designed to minimize fatigue, individual stress, and interpersonal tension, such as allocated "time off," will be critical to preserving and promoting the behavioral health of astronaut personnel.

Astronauts also train in analog environments (including Antarctica, the underwater NASA Extreme Environment Mission Operations [NEEMO], and winter and mountain survival programs). The training environments include both physical capabilities and problem resolution involving group dynamics.

The National Research Council (NRC) Committee for the Decadal Survey on Biological and Physical Sciences in Space noted that "continued research is required to identify individual, interpersonal, cultural, and environmental determinants of crew cohesion, crew performance, and ground-crew interaction" (NRC, 2011, p. 89). Such research may lead to greater specificity of behavioral health standards. The NASA Human Research Program is working on addressing research gaps in this area (see Box 3-1).

BOX 3-1
NASA Human Research Program's Research Gaps
Behavioral Health and Performance Risks

BMed 1: What are the most effective methods to enhance behavioral health and prevent decrements before, during, and after spaceflight missions?

BMed 2: What are the most effective methods to predict, detect, and assess decrements in behavioral health (which may negatively affect performance) before, during, and after spaceflight missions?
BMed 3: What aspects, if any, of cognitive performance change in-flight? If there are changes, do they persist post mission? If so, for how long?
BMed 4: What are the most effective methods for detecting and assessing cognitive performance during exploration missions?
BMed 5: What individual characteristics predict successful adaptation and performance in an isolated, confined, and extreme environment, especially for long-duration missions?
BMed 6: What are the most effective methods for treating the individual to remedy behavioral health problems during spaceflight missions (including behavioral health meds)?
BMed 7: What are the most effective methods for modifying the environment to prevent and remedy behavioral health problems during spaceflight missions?
BMed 8: How do family, friends, and colleagues affect astronauts' behavioral health and performance before, during, and after spaceflight?

Team Gap Risks

Team Gap 1: We need to understand the key threats, indicators, and life cycle of the team for autonomous, long duration, and/or distance exploration missions.
Team Gap 2: We need to identify a set of validated measures, based on the key indicators of team function, to effectively monitor and measure team health and performance fluctuations during autonomous, long duration, and/or distance exploration missions.
Team Gap 3: We need to identify a set of countermeasures to support team function for all phases of autonomous, long duration, and/or distance exploration missions.
Team Gap 4: We need to identify psychological measures that can be used to select individuals most likely to maintain team function for autonomous, long duration, and/or distance exploration missions.
Team Gap 5: We need to identify validated ground-based training methods that can be both preparatory and continuing to maintain team function in autonomous, long duration, and/or distance exploration missions.
Team Gap 6: We need to identify methods to support and enable multiple distributed teams to manage shifting levels of autonomy during long duration, and/or distance exploration missions.
Team Gap 8: We need to identify psychological and psychosocial factors, measures, and combinations thereof that can be used to compose highly effective crews for autonomous, long duration, and/or distance exploration missions.
Team Gap 9: We need to identify spaceflight acceptable thresholds (or ranges) of team function, based on key indicators, for autonomous, long duration, and/or distance exploration missions.

Sleep and Cognition Risks

Sleep Gap 1: We need to identify a set of validated and minimally obtrusive tools to monitor and measure sleep-wake activity and associated performance changes for spaceflight.
Sleep Gap 2: We need to understand the contribution of sleep loss, circadian desynchronization, extended wakefulness and work overload, on individual and team behavioral health and performance (including operational performance), for spaceflight.
Sleep Gap 4: We need to identify indicators of individual vulnerabilities and resiliencies to sleep loss and circadian rhythm disruption, to aid with individualized countermeasure regimens, for autonomous, long duration, and/or distance exploration missions.
Sleep Gap 5: We need to identify environmental specifications and operational regimens for using light to prevent and mitigate health and performance decrements due to sleep, circadian, and neurobehavioral disruption, for flight, surface and ground crews, during all phases of spaceflight operations.
Sleep Gap 6: We need to identify how individual crew members can most effectively and safely use medications to promote sleep, alertness, and circadian entrainment, as needed during all phases of spaceflight operations.
Sleep Gap 8: We need to develop individualized scheduling tools that predict the effects of sleep-wake cycles, light and other countermeasures on performance, and can be used to identify optimal (and vulnerable) performance periods during spaceflight.
Sleep Gap 9: We need to identify an integrated, individualized suite of countermeasures and protocols for implementing these countermeasures to prevent and/or treat chronic partial sleep loss, work overload, and/or circadian shifting, in spaceflight.
Sleep Gap 10: We need to identify the spaceflight environmental and mission factors that contribute to sleep decrements and circadian misalignment, and their acceptable levels of risk.

NOTE: Team Gap 7 merged with Team Gap 3. Sleep Gap 3 is closed. Sleep Gap 7 merged with Sleep Gap 2.
SOURCE: NASA, 2014c.

Individual Variation Issues

Several studies have pointed to phenotypic and genotypic variations to sleep disruption and its behavioral consequences (Van Dongen et al., 2004; Landholt, 2008; Kuna et al., 2012), suggesting the identification of predictive biomarkers (Czeisler, 2011) that could prove useful in identifying astronauts likely to experience sleep-related performance decrements and for managing sleep-wake regulation during exploration spaceflight (Goel and Dinges, 2012).

Individual characteristics identified as predictors of social compatibility in analog studies and surveys of astronaut personnel include low extraversion and high introversion (Palinkas and Suedfeld, 2008), high positive instrumentality (goal oriented, active, self-confident) and expressiveness (kind, aware of others' feelings), and low negative instrumentality (arrogant, hostile, boastful, egotistical) and communion (self-subordinating, subservient, unassertive) (McFadden et al., 1994; Rose et al., 1994).

Crew heterogeneity with respect to social (e.g., age, sex, cultural background), psychological (need for achievement, aggressiveness, autonomy), and other characteristics (e.g., interest in leisure activities) also predicts crew cohesion and conflict, Team Gap decision making, and response to crisis (NRC, 1998, 2011; Kanas and Manzey, 2008; Kanas et al., 2009).

Health Standards and Risk Profile

According to NASA's Human Research Roadmap, the research ratings for risk of adverse behavioral conditions and psychiatric disorders has been determined to be "Controlled" for the ISS and lunar missions, "Acceptable" for near-Earth asteroid missions, and "Unacceptable" for the Mars design reference mission (see Table 2-2) (NASA, 2013). The risk of fatigue-related performance errors is considered "Controlled" for all four design reference missions. The research rating for the third risk related to Team Gap issues is currently listed as "Controlled" for the lunar design reference mission and "Acceptable" for the ISS, near-Earth asteroid, and Mars missions. For a mission to a near-Earth asteroid or Mars, mission duration and distance from Earth are among the major stressors. A NASA health standard for behavioral health and performance has been developed that addresses all of these risks (see Box 2-1).

Cognitive and psychiatric assessments and screening criteria are used in the initial astronaut selection process and in the annual astronaut recertification medical examinations. Furthermore, processes have been developed by the ISS partner agencies that detail neurocognitive and behavioral health baseline assessments and follow-up.

Issues for Long Duration and Exploration Spaceflights

Few established policies regarding limits to long-duration isolation and confinement exist. The National Science Foundation's Division of

Polar Programs requires that all candidates for winter-over duty at Antarctica's Amundsen-Scott South Pole and McMurdo stations undergo a psychiatric evaluation conducted by a civilian contractor. The U.S. Antarctic Program also places limits on the number of continuous seasons (e.g., summer and winter) a person can stay at the same station, essentially mandating that personnel either spend a subsequent year or part of a year at another station or go home for a season prior to returning.

Astronauts need to be fully informed that individual and interpersonal issues that typically are of little concern on Earth or during short duration missions have the potential to become clinically and operationally significant in conditions of prolonged isolation and confinement.

While more is known about controlling the risk of fatigue-related performance errors for shorter flights, much remains to be learned about the chronic stressors of remote missions. Those missions will require autonomous operations that will change the ways crews conduct tasks, leaving them potentially more vulnerable to fatigue-related performance errors and more open to challenges regarding crew dynamics and cohesion.

Ethics issues raised in considering these risks include appropriateness of crew selection criteria (e.g., genetic, sex, and cultural differences), the necessity of informed decision making, and challenges and tensions between astronaut privacy and the need to continuously learn about the impact of spaceflight on behavior and performance. However, as with other areas of human performance, there remains a great detail of uncertainty in the field of clinical psychology related to the ability to predict individual behavior or group behavior on the basis of individual characteristics, especially in isolated and confined extreme environments (Palinkas and Suedfeld, 2008). Clinical assessments have proved to be more successful at "screening out" individuals who are unsuited to living and working in such environments; however, they have been shown to have less utility in "screening in" individuals who are best qualified for such assignments (Grant et al., 2007). Similarly, the adaptation of countermeasures found to be effective on Earth or on the ISS to long duration and exploration space missions remains controversial given their limited external validity due to context-specific influences (Shepanek, 2005). As with other health issues, establishing ethics principles for these missions must take into consideration the feasibility and desirability of doing so in the face of uncertainty.

BONE DEMINERALIZATION

Bone demineralization during exposure to microgravity illustrates the multiple parameters of a set of related health risks; how these risks are assessed, studied, and managed by NASA; countermeasure development; and interaction with engineering systems.

Overview and Risk Identification

Bone demineralization in microgravity is a well-recognized and well-studied phenomenon that still is not completely understood. It overlaps significantly with areas of highly active research for terrestrial medicine, but it is not fully known in what ways microgravity-induced bone loss might be similar to, or different from, osteoporosis.

Bone normally remodels in response to physical load, becoming stronger where that load is stronger and weaker where there is less load. Approximately a 10th of the human skeleton is normally renewed annually (Sibonga et al., 2008b). In microgravity, bone and muscle experience less loading, and bone mineral density loss rates of approximately 1 to 1.5 percent *per month* have been observed. This is compared to the 2 to 3 percent loss in bone mineral density *per year* in postmenopausal females during the first decade after the onset of menopause, which is considered a time of rapid bone loss (Sibonga et al., 2008b). Bone loss is primarily due to increased resorption of bone with subsequent release of calcium into the bloodstream and excretion by the kidneys.

Risk Management: Countermeasures and Research

Because of significant overlap with important clinical questions for terrestrial health care (e.g., osteoporosis and metabolic bone diseases), human and animal models, monitoring methods (e.g., biochemical markers, bone ultrasound, and dual-energy x-ray absorptiometry scan), and a variety of countermeasures (pharmacologic, dietary, and exercise) have been developed and are being implemented for astronauts (Sibonga et al., 2008a,b). However, much remains to be learned about the effects of microgravity, particularly for long duration exposures.

Bone loss in microgravity is exacerbated by dietary factors. Documented undernutrition ("space anorexia"), lack of adequate vitamin D, a diet relatively high in sodium (which increases kidney urine calcium concentration), and the particular relative composition of amino acids in the diet have all been implicated as contributing factors and are all amenable to careful nutritional adjustment (Smith et al., 2012).

Potential countermeasures include a combination of aerobic and resistive exercise, nutrition, vitamin D supplementation, and pharmacologic agents such as bisphosphonates. Most recent studies in a small number of astronauts on the ISS have combined optimal nutrition (adequate caloric intake, low dietary sodium, and appropriate amino acid balances) and vitamin D supplementation with an aggressive exercise program (Smith et al., 2012). These relatively successful countermeasures increase the rate of new bone formation, so bone mass is maintained, but there is increased bone-cell turnover. Pharmacologic agents that are used to combat bone loss associated with menopause (in women) and aging (in both men and women) have been studied to a very limited extent in the space environment. Both bone mass and bone architecture contribute to bone strength and fracture risk. New bone that is deposited as a result of countermeasures appears to exhibit altered architecture compared with the bone that has been lost (Sibonga et al., 2008b). Moreover, the increased calcium turnover may increase the risk of kidney stones.[2] Drinking more water dilutes the calcium concentration and decreases this risk.

Thus, demineralization poses risk to an individual astronaut during and after the mission, and risk to the mission if bone loss leads to fracture. A combination of countermeasures was tested on 13 crew members during several ISS missions and was associated with increased formation of new bone (without a corresponding decrease in resorption) and overall maintenance of bone mass (Smith et al., 2012).

Although multiple knowledge gaps remain, one of particular interest is development of an easy, noninvasive way to monitor bone health in microgravity. Research gaps are identified in the NASA Human Research Program's evidence reports on this health concern (Sibonga et al., 2008a,b).

Individual Variation Issues

The significant individual variation in bone demineralization is so wide as to obscure differences due to biological sex characteristics or any

[2]An unexpected environmental support system hazard occurred during the summer of 2009 when a new type of urine processing assembly (a critical unit that allows water reclamation) broke down shortly after being brought on line in the ISS. The cause was eventually traced to precipitation of calcium from urine in a mechanism strikingly similar to the biological process of renal stone formation (Smith et al., 2012).

other associated predisposing factors (Ploutz-Snyder, 2013). No predictive or explanatory model exists for understanding this range in individual variation.

Health Standards and Risk Profile

The risk of fracture due to bone demineralization is classified as "Controlled" for missions to the ISS or to a near-Earth asteroid, and "Acceptable" for lunar and Mars design reference missions (see Table 2-2). The risk of early-onset osteoporosis due to spaceflight is ranked "Acceptable" for all four design reference missions (see Table 2-2). A NASA health standard for bone mineral loss has been developed that addresses both of these risks (see Box 2-1).

Issues for Long Duration and Exploration Spaceflights

An unknown variable is whether the rate of bone loss for astronauts during prolonged missions will remain consistent with rates observed in shorter (6 months or less) missions, whether it will plateau, or whether it may accelerate as space environment exposure lengthens. The long-term effects of bone loss are relatively well known for Earth-based populations. Both menopause-related and senile osteoporosis are associated with an increased risk of fracture (primarily hip, vertebral, and wrist). While bone demineralization theoretically could increase the risk of actual fracture during a long duration spaceflight, this has never occurred during a mission, although obviously, such an occurrence could jeopardize successful completion of the mission or lead to morbidity affecting astronaut well-being after the mission.

Bone density is closely monitored in the active astronaut corps as well as in those astronauts who participate in the Longitudinal Study of Astronaut Health. During the ISS flights, astronauts exercise extensively, and their diet is tailored for maintaining bone density. After each flight, bone scans are done to examine the degree of bone loss, and protocols (both pharmaceutical and physical) are recommended for bone recovery. Postflight rehabilitation aims to return bone mass levels to preflight baselines (NASA, 2007).

Long duration and exploration spaceflights with landings, such as a mission to Mars, could expose astronauts to additional risks beyond those associated with a long stay on the ISS. Such an exploration-class mission, as currently envisioned, would include a period of exposure to microgravity, the physical stress of landing, exposure to and activities in the fractional gravity of Mars (0.375 of the gravity of Earth; NASA, 2014d), physical stress during takeoff from Mars, exposure to microgravity again, and finally physical stress during landing on Earth. Essentially nothing is known about how exposure to fractional gravity environment on Mars might influence the time course of microgravity-induced bone loss or fracture during the mission.

Ethics issues raised in consideration of this health risk include potential impact on crew selection, responsibilities for long-term monitoring and health of astronauts, balancing the unknowns regarding this risk with other risks and benefits, and ensuring informed decision making.

RADIATION EXPOSURE

For space missions in low Earth orbit (LEO), the major source of radiation exposure results from solar storms. For exploration-class missions beyond LEO, it is the exposure to galactic cosmic radiation that is the significant health concern for both acute and long-term health consequences. High levels of radiation exposure (e.g., during solar storms) can lead to acute effects, including fatigue, nausea, and vomiting. Chronic exposure increases the risk of cancer, tissue degeneration, development of cataracts, and potential effects on the central nervous system, cardiovascular system, immune function, and vision.

Overview and Risk Identification

NASA recognized the potential radiation risk from the beginning of its efforts to send astronauts into space. An extensive radiation research program has been under way and input has been received from a number of independent organizations (see Box 3-2).

BOX 3-2
Overview of Selected Reports on Radiation Exposure Limits

National Research Council (NRC, 1967), ***Radiobiological Factors in Manned Space Flight.*** The working group focused on identifying "immediate or early performance decrement (early responses) occurring within a few hours to 1 month of major exposure;" "progressively increasing performance decrement or serious loss of performance over longer period of flight as a result of an accumulating exposure (progressive injury to the blood-forming system);" and "probability of late radiation response" (p. 244).

National Research Council (1970), *Radiation Protection Guides and Constraints for Space-Mission and Vehicle-Design Studies Involving Nuclear Systems*. The NRC committee noted that the risk-benefit decisions depend on "a wide range of general and specific scientific and subjective judgments and should be the responsibility of those most informed about the aims and goals of the nation's space program" (p. 3). The report noted the value of specific limits to spacecraft design and mission planning. The final recommendations of the NRC committee included a set of short-term guidelines to limit acute and degenerative effects, along with the concept of a "primary reference risk" to limit the exposure to that which "corresponds to an added probability of radiation-induced neoplasia over a period of about 20 years that is equal to the natural probability for the specific population at risk" (p.16). They estimated this exposure to be 400 REM, which they believed corresponded to about a 2.3 percent risk of developing cancer. However, they noted that acceptance of a higher risk for planetary missions than for space station missions "would seem both realistic and practical" (p. 16).

National Council on Radiation Protection and Measurements (NCRP) Report 98, *Guidance on Radiation Received in Space Activities* (1989). In addition to recommending short-term limits to minimize acute effects, this study noted that cancer was the principal risk, and career limits were set to limit the risk of fatal cancer. The report limited the risk to 3 percent excess lifetime cancer mortality, based on comparison with other hazardous occupations. The study was also the first to take age and sex into account.

NCRP Report 132, *Radiation Protection Guidance for Activities in Low-Earth Orbit* (2000). NCRP 132 questioned the use of mortality data from hazardous occupations as the basis for space-related radiation limits, but endorsed the 3 percent lifetime risk of cancer death, as consistent with guidelines for terrestrial radiation workers. The report also briefly but explicitly addressed the challenge of uncertainties in quantifying radiation limits: "It is well known that risk estimation is a difficult field in which there are many sources of potential error and therefore uncertainty.... Given the magnitude of these uncertainties and the problems of dose specification estimates or risk on which dose limits for astronauts are based should be recognized as very conservative and possibly subject to modified values when more precise information becomes available" (pp. 146-147). Report 132 clarified that the NCRP had only been tasked to assess risk to low Earth orbit (LEO) missions: "In NCRP Report No. 98 (NCRP, 1989) a section on mission scenarios with estimates of radiation exposure during missions to the moon and Mars was included. Perhaps because of that inclusion, some have assumed that the guidance on dose limits in that report applied not only to missions in LEO but to all space missions. That was not the intention since the guidance provided was limited to exposures in LEO. In this Report, the guidance is also only intended to be applied for radiation exposures incurred during missions in LEO. Further NCRP reports will deal with other space situations" (pp. 2-3).

NCRP Report 153, *Information Needed to Make Radiation Protection Recommendations for Space Missions Beyond Low-Earth Orbit* (2006). "The purpose of this Report is to identify and describe information needed to make radiation protection recommendations for space missions beyond low-Earth orbit (LEO). Current space radiation guidelines pertain only to missions in LEO and are not considered relevant for mission beyond LEO" (NCRP, 2006, p. 1).

NRC (2012a), *Technical Evaluation of the NASA Model for Cancer Risk to Astronauts Due to Space Radiation.* The NRC committee did not address the underlying assumption of whether the permissible exposure limits were appropriate, rather they looked in detail at the constituent elements of the model used to calculate the risk. They found the model to be consistent with general radiation community approaches to quantify the radiation risk.

SOURCES: NRC, 1967, 1970, 2012a; NCRP, 1989, 2000, 2006.

Current career limits in the NASA standards are designed to keep the astronaut's lifetime risk of exposure-induced death from cancer to no more than 3 percent (i.e., limiting the additional risk of cancer posed by mission-related radiation exposure). The risk and lifetime exposure limit can be explained as follows: If 100 astronauts were exposed to the upper bounds of the radiation limits, 3 would die of cancer attributable to that exposure. It would be anticipated that life expectancy for astronauts with radiation-induced cancer would be reduced by an average of 12 to 16 years compared to those without radiation-induced cancer (NASA, 2007).

In 1990, NASA agreed to accept the recommendation of NCRP Report 98 to limit the lifetime risk to astronauts to 3 percent and formalized it in the 1995 NASA Health Standards (NASA, 1995). The 1995 standards stated the permissible exposure limit in terms of sex- and age-dependent, dose-equivalent limits expressed in REMs,[3] which were developed to be consistent with the 3 percent risk of exposure-induced death (REID) (e.g., 237.5 REM for a 35-year-old male, 177.5 REM for a 35-year-old female) (NASA, 1995). These estimated limits did not include a margin to represent the uncertainty in the calculations. As a result

[3]REM (rad equivalent man) was used as a unit of dose equivalent in the 1995 NASA standard, but NASA now uses the sievert (Sv), the international standard for dose equivalent (NRC, 2012a). 1 REM is equal to 0.01 Sv.

of NCRP Report 132, published in 2000, NASA proposed to add a 95 percent confidence interval to the 3 percent limit. The current radiation permissible exposure limits with the 95 percent confidence interval was formally accepted in March 2007 (NASA, 2007) (see Box 2-1). The procedure used to estimate the confidence interval is based on extrapolation from a wide range of physical and biological research, much of it conducted by NASA in a peer-reviewed research program open to the wider radiation research community. The model and the underlying methodology and assumptions are periodically reviewed by independent committees, most recently by the NRC Space Studies Board (NRC, 2012a).

Risk Management: Research, Exposure Limits, Monitoring, and Countermeasures

NASA has undertaken a research program to quantify the estimates of the risk of radiation-related cancers and other acute and chronic adverse health effects (Cucinotta and Durante, 2009; Cucinotta et al., 2009; Huff and Cucinotta, 2009; Wu et al., 2009). Progress based on a peer-reviewed, science-based approach has significantly reduced the uncertainties associated with the estimates of the risk, and the long-term program goal is to reduce it further.

Radiation levels are monitored in real time on the ISS by a suite of sensors distributed throughout the station. The reports from these sensors are used to estimate astronaut exposure throughout the mission. When the rate of exposure exceeds threshold levels, mission control is informed. For short periods, when radiation levels are significantly above normal, astronauts may be instructed to stay in better shielded areas of the ISS. In addition, each astronaut has a personal dosimeter that is read after the mission by the NASA Space Radiation Analysis Group to confirm and record exposure. For exploration missions, research is under way to develop real-time personal dosimeters.

NASA's Space Radiation Analysis Group works with the National Oceanic and Atmospheric Administration's Space Weather Prediction Center to monitor the radiation levels near Earth and monitor for evidence of solar storms. The center issues alerts when a radiation storm is under way, and the Space Radiation Analysis Group alerts ISS mission operations.

Countermeasures for radiation exposure are limited. The primary countermeasure is shielding built into the spacecraft. However, due to the highly penetrating nature of galactic cosmic rays, shielding is only marginally effective at reasonable thicknesses; increasing the thickness of the shielding adds substantial mass with minimal additional reduction in exposure (NRC, 2008). Innovative approaches to reduce the risks also are being investigated, including pharmacological countermeasures. To date, these methods have not been shown to be effective.

Thirty-day exposure limits have been established to limit the potential for acute health outcomes due to radiation exposure, and operational processes and procedures are implemented to ensure that astronauts do not approach these limits. The greatest likely source of acute exposure is solar storms. Providing access to modest shielding and a system for timely warning of a pending or ongoing storm are critical elements of a risk management strategy to mitigate this risk (NRC, 2008). Pre-mission planning includes estimates of the radiation exposure to minimize the risk that the crew will exceed the career limits (with consideration of the crew member's prior exposure).

Radiation exposures also may include risks of chronic, degenerative effects that could affect health during a lengthy mission. Only recently has quantitative progress been made in understanding these risks, particularly those affecting the central nervous and circulatory systems (Cucinotta et al., 2013b). An ongoing research effort is under way to improve this understanding.

Health Standards and Risk Profile

NASA has long recognized the threat of radiation, and has established space permissible exposure limits to protect its astronauts (see Box 2-1). The three tiers of radiation exposure limits in the NASA standards address: career limits, designed to limit the lifetime risk of exposure-induced death from cancer; short-term limits, designed to limit the risk of acute affects; and operational processes and procedures which emphasize that "in-flight radiation exposures shall be maintained using the 'as low as reasonably achievable' (ALARA) principle" (see Box 2-1; NASA, 2007, p. 20).

Radiation limits are significant in planning for long duration and exploration missions. Existing radiation standards, developed for the Space Shuttle and the ISS, would limit missions to durations of 150 to 250 days (Cucinotta et al., 2013a). The radiation standards are written to apply to all NASA human spaceflight missions and are not developed for any specific program (NASA, 2007). However, while some of the existing programs, such as the Space Shuttle and the ISS programs, can be conducted within the standards, these standards impose potential limitations on long-duration missions (e.g., a 1-year stay on the ISS) or missions with architectures and objectives outside of LEO. The standards are based on recommendations to NASA from NCRP in a report that focused on LEO missions (NCRP, 2000). The NCRP report noted, "In this Report, the guidance is also only intended to be applied for radiation exposures incurred during missions in LEO" (NCRP, 2000, pp. 2-3).

Individual Variation Issues

A specific individual's cancer risks vary by age, sex, and race, as well as genetic factors and life-style choices. National databases allow calculation by age, sex, and other factors, emphasizing the role of individual variability. NASA's standards limit the *additional* risk of cancer posed by radiation exposure, not the total risk of dying from cancer. If one assumes a generic baseline lifetime risk of 23 percent for males and 19 percent for females (ACS, 2013), and leaves aside any differences in individual susceptibility, then male astronauts exposed to the 3 percent radiation exposure limit would have an estimated lifetime risk of death from cancer of 26 percent, while female astronauts would have an estimated risk of 22 percent.

In general, for females the effective-dose radiation exposure is about three quarters that of males before reaching career limits (NASA, 2007). Females, on average, are more susceptible to radiation-induced cancer, in part because of the baseline higher risk of breast cancer in females; body size also is a factor as individuals with smaller builds have less body self-shielding. However, an individual's risk of cancer is dominated by a number of factors (including genetic susceptibility, lifestyle choices, and prior exposures, both natural and medical), most of which may be highly individualized or unknown and, therefore, not currently possible to factor in calculations of cancer risk.

Issues for Long Duration and Exploration Spaceflights

As noted above, the number of days that astronauts could be exposed and stay within the current permissible exposure limits (safe days) will depend on assumptions about shielding and the radiation environment. The exposure varies with the 11-year solar cycle and with the number and intensity of solar storms (Cucinotta et al., 2013a). The existing radiation exposure limits would limit long duration or exploration missions to

150 to 250 days (Cucinotta et al., 2013a). The estimated time in the zero-g interplanetary environment ("deep space")[4] for a Mars mission with current propulsion systems is estimated at 400 to 600 days (NASA, 2009). The design reference missions to Mars that are being considered include those that would include a long stay on the Martian surface (500 or more days) with a shorter stay in deep space, as well as those that would have a longer deep space duration and shorter Mars surface stays, nominally 30 to 90 days (NASA, 2009). Technology research into alternative propulsion systems could significantly reduce the Mars transit time (NRC, 2012b).

Other factors also could affect the number of safe days in space with regard to radiation exposure. For example, noting that astronauts, in general, are healthier than the average U.S. population, NASA recently calculated the number of safe days in space using the data on cancer risks for non-smoking, normal weight Americans (see Table 3-1). This change in the population data increases the estimates of safe days in space by 30 to 90 percent, depending on an astronaut's age and sex (Cucinotta et al., 2013a).

The NASA current precautionary decision to set exposure limits to protect astronauts with 95 percent probability (given the level of uncertainty) itself limits the number of safe days in space. If a less precautionary approach were taken by NASA using a lower confidence interval, the number of allowable days in space would be greater; alternatively, if greater precaution were exercised by using a higher confidence interval, fewer days would be permissible. More precise information about the dose of radiation-inducing fatal cancers could also change the number of permissible days in space that would not exceed the health standard.

Ethics issues raised in the consideration of radiation exposure include informed decision making by astronauts, individual variation and other factors that might impact crew selection, and assessing and balancing the unknown risks of radiation exposure with other risks and benefits of the mission. While NASA has analyzed health risks of radiation based on age and sex (see Table 3-1), other factors influencing individual variation and radiation risk must also be considered. Existing data on such variation should be factored into analyses, and new data should be collected to better predict individual risk from exposure to radiation.

[4]Calculations regarding radiation exposure factor in the higher exposures to radiation in interplanetary space outside Earth's magnetosphere and the relatively lower exposure levels on a planetary or lunar surface.

TABLE 3-1 Estimates of Safe Days in Deep Space[a]

	Average solar minimum GCR		Average solar maximum GCR and one significant solar storm (similar to that which occurred in August 1972)	
Age at Exposure (years)	NASA 2012 U.S. Average Population	NASA 2012 Never-smokers	NASA 2012 U.S. Average Population	NASA 2012 Never-smokers
Males				
35	209 (205)	271 (256)	306 (357)	395 (458)
45	232 (227)	308 (291)	344 (397)	456 (526)
55	274 (256)	351 (335)	367 (460)	500 (615)
Females				
35	106 (95)	187 (180)	144 (187)	276 (325)
45	139 (125)	227 (212)	187 (232)	319 (394)
55	161 (159)	277 (246)	227 (282)	383 (472)

NOTE: Solar minimum is a 2- to 3-year period of low solar activity in the 10- to 11-year solar cycle. Solar maximum is a corresponding 5- to 7-year period of enhanced solar activity. Galactic cosmic radiation is at a peak during solar minimum and somewhat reduced during solar maximum. Values in parentheses for solar minimum are for the deep solar minimum of 2009. Values in parentheses for solar maximum are for the case where a storm shelter is available to reduce the solar storm exposure to a negligible amount. GCR = galactic cosmic radiation; REID = risk of exposure-induced death.

[a]Safe days in deep space (zero-g interplanetary environment) is a concept defined as the maximum number of days with 95 percent confidence interval to be below the NASA 3 percent REID limit, assuming nominal shielding of 20 g/cm^2 aluminum.

SOURCE: Adapted from Cucinotta et al., 2013a.

DISCUSSION: Because the highly penetrating GCR dominates the effective dose, and because GCR is at a peak at solar minimum, the safe days are lowest at solar minimum. The deep solar minimum of 2009 had a significantly higher GCR than previously experienced, thus further reducing the safe days. There are more safe days near solar maximum because the GCR is reduced relative to solar minimum, and because solar storm radiation is significantly attenuated by nominal shielding. If a storm shelter (a smaller volume within the vehicle with additional shielding) is available to the astronauts, the exposure from a solar storm can be reduced to a negligible amount, so the total exposure would be from the GCR alone. This could further increase the number of safe days at solar maximum, assuming adequate solar storm warning.

SUMMARY

The examples provided in this chapter illustrate the range of issues that are faced as NASA makes decisions about health standards for long duration and exploration spaceflights. The potential short- and long-term health impacts encompass many systems of the human body as well as behavior and performance issues. Decisions about health standards are complicated by the depth of uncertainty regarding what will happen with extended stays in space, high exposures to galactic cosmic radiation, and other risks and challenges. Additionally, data are minimal or non-existent for variations in individual susceptibility based on factors such as ethnicity, sex, age, etc. A major challenge is the lack of information about the interaction of risks and the extent to which these interactions alter the overall level of risk. The following chapter outlines the ethics principles and responsibilities that can be brought to bear on decisions regarding health risks.

REFERENCES

ACS (American Cancer Society). 2013. *Lifetime risk of developing or dying from cancer*. http://www.cancer.org/cancer/cancerbasics/lifetime-probability-of-developing-or-dying-from-cancer (accessed January 15, 2014).

Aldrin, B. 1973. *Return to Earth*. New York: Random House.

Alexander, D. J., C. R. Gibson, D. R. Hamilton, S. M. C. Lee, T. H. Mader, C. Otto, C. M. Oubre, A. F. Pass, S. H. Platts, J. M. Scott, S. M. Smith, M. B. Stenger, C. M. Westby, and S. B. Zanello. 2012. *Evidence report: Risk of spaceflight-induced intracranial hypertension and vision alterations*. http://humanresearchroadmap.nasa.gov/Evidence/reports/VIIP.pdf (accessed November 8, 2013).

Basner, M., D. F. Dinges, D. Mollicone, A. Ecker, C. W. Jones, E. C. Hyder, A. D. Antonio, I. Savelev, K. Kan, N. Goel, B. V. Morukov, and J. P. Sutton. 2013. Mars 320-d mission simulation reveals protracted crew hypokinesis and alterations of sleep duration and timing. *Proceedings of the National Academy of Sciences of the United States of America* 110(7):2635-2640.

Behnken, R., M. Barratt, S. Walker, and P. Whitson. 2013. *Presentation to the Institute of Medicine, Ethics Principles and Guidelines for Health Standards for Long Duration and Exploration Spaceflights: Astronaut Office*. PowerPoint presented at the second meeting of the Committee on Ethics Principles and Guidelines for Health Standards for Long Duration and Exploration Spaceflights. Washington, DC, July 25. http://www.iom.edu/~/media/Files/Activity%20Files/Research/HealthStandardsSpaceflight/2013-JUL-25/Panel%202%20Astronaut%20Corp%20Final%20IOM_presentation final2.pdf (accessed November 8, 2013).

Billica, R. 2000. *Inflight medical events for U.S. astronauts during space shuttle programs STS-1 through STS-89, April 1981–January 1998*. Presentation to the Institute of Medicine Committee on Creating a Vision for Space Medicine During Travel Beyond Earth Orbit. NASA Johnson Space Center, Houston, February 22. As cited in: IOM, 2001. *Safe passage: Astronaut care for exploration missions*. Washington, DC: National Academy Press.

Burrough, B. 1998. *Dragonfly: NASA and the crisis aboard Mir*. New York: HarperCollins.

Cucinotta, F. A., and M. Durante. 2009. *Evidence report: Risk of radiation carcinogenesis*. http://humanresearchroadmap.nasa.gov/Evidence/reports/Carcinogenesis.pdf (accessed January 8, 2014).

Cucinotta, F. A., H. Wang, and J. L. Huff. 2009. *Evidence report: Risk of acute or late central nervous system effects from radiation exposure*. http://humanresearchroadmap.nasa.gov/Evidence/reports/CNS.pdf (accessed January 8, 2014).

Cucinotta, F. A., M.-H. Y. Kim, and L. J. Chappell. 2013a. *Space radiation cancer risk projections and uncertainties—2012*. http://ston.jsc.nasa.gov/collections/TRS/_techrep/TP-2013-217375.pdf (accessed November 11, 2013).

Cucinotta, F. A., M.-H. Y. Kim, L. J. Chappell, and J. L. Huff. 2013b. How safe is safe enough? Radiation risk for a human mission to Mars. *PLoS ONE* 8(10):e74988.

Czeisler, C. A. 2011. Impact of sleepiness and sleep deficiency on public health—utility of biomarkers. *Journal of Clinical Sleep Medicine* 7(5):s6-s8.

Czeisler, C. A., J. S. Allan, S. H. Strogatz, J. M. Ronda, R. Sanchez, C. D. Rios, W. O. Freitag, G. S. Richardson, and R. E. Kronauer. 1986. Bright light resets the human circadian pacemaker independent of the timing of the sleep-wake cycle. *Science* 233(4764):667-671.

Dinges, D. F., F. Pack, K. Williams, K. A. Gillen, J. W. Powell, G. E. Ott, C. Aptowicz, and A. I. Pack. 1997. Cumulative sleepiness, mood disturbance, and psychomotor vigilance performance decrements during a week of sleep restricted to 4-5 hours per night. *Sleep* 20(4):267-277.

ESA (European Space Agency). 2011. *Mars500: Study overview*. http://www. http://www.esa.int/Our_Activities/Human_Spaceflight/Mars500/Mars500_study_overview (accessed January 8, 2014).

Flynn, C. F. 2005. An operational approach to long-duration mission behavioral health and performance factors. *Aviation, Space, and Environmental Medicine* 233(Suppl. 6):B42-B51.

Garber, S., and R. Launius. 2005. *A brief history of NASA*. http://history.nasa.gov/factsheet.htm (accessed December 4, 2013).

Goel, N., and D. F. Dinges. 2012. Predicting risk in space: Genetic markers for differential vulnerability to sleep restriction. *Acta Astronautica* 77:207-213.

Grant I., H. R. Eriksen, P. Marquis, I. J. Orre, L. Palinkas, P. Suedfeld, E. Svensen, and H. Ursin. 2007. Psychological selection of Antarctic personnel: The "SOAP" instrument. *Aviation, Space, and Environmental Medicine* 78:793-800.

Gunderson, E. K. E. (Ed.). 1974. Psychological studies in Antarctica. In *Human adaptability to Antarctic conditions.* Washington, DC: American Geophysical Union. Pp. 115-131.

Gushin, V. I., T. B. Kolinitchenko, V. A. Efimov, and C. Davies. 1996. Chapter 16: Psychological evaluation and support during exemsi. In *Advances in Space Biology and Medicine.* Vol. 5, edited by S. L. Bonting. Greenwich, CT: JAI Press. Pp. 283-295.

Huff, J. L., and F. A. Cucinotta. 2009. *Evidence report: Risk of degenerative tissue or other health effects from radiation exposure*. http://humanresearch roadmap.nasa.gov/Evidence/reports/Degen.pdf (accessed January 8, 2014).

IOM (Institute of Medicine). 2001. *Safe passage: Astronaut care for exploration missions*. Washington, DC: National Academy Press.

IOM. 2014. *Review of NASA's evidence reports on human health risks: 2013 letter report*. Washington, DC: The National Academies Press.

Kanas, N., and D. Manzey. 2008. *Space psychology and psychiatry*. 2nd ed. El Segundo, CA: Microcosm Press.

Kanas, N. A., V. P. Salnitskiy, J. E. Boyd, V. I. Gushin, D. S. Weis, S. A. Saylor, O. P. Kozerenko, and C. R. Marmar. 2007. Crewmember and mission control personnel interactions during International Space Station missions. *Aviation, Space, and Environmental Medicine* 78(6):601-607.

Kanas, N., G. Sandal, J. E. Boyd, V. I. Gushin, D. Manzey, R. North, G. R. Leon, P. Suedfeld, S. Bishop, E. R. Fiedler, N. Inoue, B. Johannes, D. J. Kealey, N. Kraft, I. Matsuzaki, D. Musson, L. A. Palinkas, V. P. Salnitskiy, W. Sipes, J. Stuster, and J. Wang. 2009. Psychology and culture during long-duration space missions. *Acta Astronautica* 64(7-8):659-677.

Kuna, S. T., G. Maislin, F. M. Pack, B. Staley, R. Hachadoorian, E. F. Coccaro, and A. I. Pack. 2012. Heritability of performance deficit accumulation during acute sleep deprivation in twins. *Sleep* 35(9):1223-1233.

Landholt, H.-P. 2008. Genotype-dependent differences in sleep, vigilance, and response to stimulants. *Current Pharmaceutical Design* 14(32):3396-3407.

Lebedev, V. 1988. *Diary of a cosmonaut: 211 days in space.* College Station, TX: Phytoresource Research, Inc.

Linenger, J. M. 2000. *Off the planet: Surviving five perilous months aboard the space station Mir*. New York: McGraw-Hill.

Mader, T. H., C. R. Gibson, A. F. Pass, L. A. Kramer, A. G. Lee, J. Fogarty, W. J. Tarver, J. P. Dervay, D. R. Hamilton, A. Sargsyan, J. L. Phillips, D. Tran, W. Lipsky, J. Choi, C. Stern, R. Kuyumjian, and J. D. Polk. 2011. Optic disk edema, globe flattening, choroidal folds, and hyperopic shifts observed in astronauts after long-duration space flight. *Ophthalmology* 18(10):2058-2069.

Mader, T. H., C. R. Gibson, A. F. Pass, A. G. Lee, H. E. Killer, H.-C. Hansen, J. P. Dervay, M. R. Barratt, W. J. Tarver, A. E. Sargsyan, L. A. Kramer, R. Riascos, D. G. Bedi, and D. R. Pettit. 2013. Optic disk edema in an astronaut after repeat long-duration space flight. *Journal of Neuro-Ophthalmology* 33:249-255.

McFadden, T. J., R. L. Helmreich, R. M. Rose, and L. F. Fogg. 1994. Predicting astronauts' effectiveness: A multivariate approach. *Aviation, Space, and Environmental Medicine* 65(10):904-909.

Molloy, M. W., and R. A. Petrone. 2013. *Apollo 12 command and service module (CSM)*. http://nssdc.gsfc.nasa.gov/nmc/spacecraftDisplay.do?id=1969-099A (accessed December 4, 2013).

NASA (National Aeronautics and Space Administration). 1995. *Man-systems integration standards. Revision B*. NASA-STD-3000. https://standards.nasa.gov/documents/viewdoc/3314902/3314902 (accessed January 8, 2014).

NASA. 2007. *NASA space flight human system standard. Volume 1: Crew health*. NASA-STD-3001. https://standards.nasa.gov/documents/detail/3315622 (accessed November 8, 2013).

NASA. 2009. *Human exploration of Mars: Design reference architecture 5.0*. NASA/SP-2009-566. http://www.nasa.gov/pdf/373665main_NASA-SP-2009-566.pdf (accessed January 8, 2014).

NASA. 2013. *Human research program requirements document, Revision F*. HRP-47052. Johnson Space Center. http://www.nasa.gov/pdf/579466main_Human_Research_Program_Requirements_DocumentRevF.pdf (accessed October 18, 2013).

NASA. 2014a. *Kennedy Space Center visitor complex: Astronaut memorial*. http://www.kennedyspacecenter.com/the-experience/astronaut-memorial.aspx (accessed January 13, 2014).

NASA. 2014b. *Mir Space Station*. http://history.nasa.gov/SP-4225/mir/mir.htm (accessed January 13, 2014).

NASA. 2014c. *Human Research Roadmap: Gaps*. http://humanresearchroadmap.nasa.gov/Gaps (accessed February 7, 2014).

NASA. 2014d. *All about Mars*. http://mars.nasa.gov/allaboutmars/extreme/quickfacts (accessed February 19, 2014).

NCRP (National Council on Radiation Protection and Measurements). 1989. *NCRP report no. 98: Guidance on radiation received in space activities*. Bethesda, MD: NCRP.

NCRP. 2000. *NCRP report no. 132: Radiation protection guidance for activities in low-Earth orbit*. Bethesda, MD: NCRP.

NCRP. 2006. *NCRP report no. 153: Information needed to make radiation protection recommendations for space missions beyond low-Earth orbit*. Bethesda, MD: NCRP.

NRC (National Research Council). 1967. *Radiobiological factors in manned space flight*. Washington, DC: National Academy Press.

NRC. 1970. *Radiation protection guides and constraints for space-mission and vehicle-design studies involving nuclear systems*. Washington, DC: National Academy Press.

NRC. 1998. *A strategy for research in space biology and medicine into the next century*. Washington, DC: National Academy Press.

NRC. 2008. *Managing space radiation risk in the new era of space exploration*. Washington, DC: The National Academies Press.

NRC. 2011. *Recapturing a future for space exploration: Life and physical sciences research for a new era*. Washington, DC: The National Academies Press.

NRC. 2012a. *Technical evaluation of the NASA model for cancer risk to astronauts due to space radiation*. Washington, DC: The National Academies Press.

NRC. 2012b. *NASA space technology roadmaps and priorities: Restoring NASA's technological edge and paving the way for a new era in space*. Washington, DC: The National Academies Press.

Palinkas, L. A., and P. Suedfeld. 2008. Psychological effects of polar expeditions. *Lancet* 371(9607):153-163.

Palinkas, L. A., F. Glogower, M. Dembert, K. Hansen, and R. Smullen. 2004. Incidence of psychiatric disorders after extended residence in Antarctica. *International Journal of Circumpolar Health* 63(2):157-168.

Ploutz-Snyder, L. 2013. *Musculoskeletal working group report*. PowerPoint presented at NASA and NSBRI Present a Virtual Workshop: The Impact of Sex & Gender on Adaptation to Space. http://www.nasa.gov/sites/default/files/files/Ploutz-Snyder_Sex-Gender_508.pdf (accessed October 22, 2013).

Rose, R. M., L. F. Fogg, R. L. Helmreich, and T. J. McFadden. 1994. Psychological predictors of astronaut effectiveness. *Aviation, Space, and Environmental Medicine* 65(10):910-915.

Rosnet, E., C. Le Scanff, and M. Sagal. 2000. How self-image and personality affect performance in an isolated environment. *Environment and Behavior* 32(1):18-31.

Sandal, G. M. 2001. Crew tension during a space station simulation. *Environment and Behavior* 33(1):134-150.

Sandal, G. M., R. J. Vaernes, T. Bergan, M. Warncke, and H. Ursin. 1996. Psychological reactions during polar expeditions and isolation in hyperbaric chambers. *Aviation, Space, and Environmental Medicine* 67(3):227-234.

Sandal, G. M., I. M. Endresen, R. Vaernes, and H. Ursin. 1999. Personality and coping strategies during submarine missions. *Military Psychology* 11(4):381-404.

Schmidt, L. L., K. Keeton, K. J. Slack, L. B. Leveton, and C. Shea. 2009. *Evidence report: Risk of performance errors due to poor Team Gap cohesion and performance, inadequate selection/Team Gap composition, inadequate training, and poor psychosocial adaptation.* http://humanresearchroadmap.nasa.gov/Evidence/reports/Team Gap.pdf (accessed January 2, 2014).

Shepanek, M. 2005. Human behavioral research in space: Quandaries for research subjects and researchers. *Aviation, Space, and Environmental Medicine* 76(Suppl. 6):B25-B30.

Sibonga, J. D., J. A. Jones, J. G. Myers, B. E. Lewandowski, T. F. Lang, and J. H. Keyak. 2008a. *Evidence report: Risk of bone fracture.* http://humanresearchroadmap.nasa.gov/evidence/reports/Bone%20Fracture.pdf (accessed October 22, 2013).

Sibonga, J. D., L. C. Shackelford, A. LeBlanc, S. Petak, S. M. Smith, S. A. Bloomfield, J. H. Keyak, T. F. Lang, S. B. Arnaud, S. Amin, B. L. Clarke, R. Wermers, P. R. Cavanagh, A. A. Licata, S. Judex, M. B. Schaffler, and T. A. Bateman. 2008b. *Evidence report: Risk of accelerated osteoporosis.* http://humanresearchroadmap.nasa.gov/evidence/reports/osteo.pdf (accessed October 22, 2013).

Slack, K. J., C. Shea, L. B. Leveton, A. M. Whitmire, and L. L. Schmidt. 2009. *Evidence report: Risk of behavioral and psychiatric conditions.* http://humanresearchroadmap.nasa.gov/evidence/reports/BMED.pdf (accessed Ocotber 18, 2013).

Smith, S. M., M. A. Heer, L. C. Shackelford, J. D. Sibonga, L. Ploutz-Snyder, and S. R. Zwart. 2012. Benefits for bone from resistance exercise and nutrition in long-duration spaceflight: Evidence from biochemistry and densitometry. *Journal of Bone and Mineral Research* 27(9):1896-1906.

Stuster, J. 2010. *Behavioral issues associated with long-duration space expeditions: Review and analysis of astronaut journals.* http://ston.jsc.nasa.gov/collections/trs/_techrep/TM-2010-216130.pdf (accessed November 8, 2013).

Tansey, W. A., J. M. Wilson, and K. E. Schaefer. 1979. Analysis of health data from 10 years of Polaris submarine patrols. *Undersea Biomedical Research* 6(Suppl.):s217-s246.

Thomas, T. L., T. I. Hooper, M. Camarca, J. Murray, D. Sack, D. Molé, R. T. Spiro, W. G. Horn, and F. C. Garland. 2000. A method for monitoring the health of US Navy submarine crewmembers during periods of isolation. *Aviation, Space, and Environmental Medicine* 71(7):699-705.

Van Dongen, H. P. A., M. D. Baynard, G. Maislin, and D. F. Dinges. 2004. Systematic interindividual differences in neurobehavioral impairment from sleep loss: Evidence of trait-like differential vulnerability. *Sleep* 27(3):423-433.

Watkins, S. D., and Y. R. Barr. 2010. *Papilledema summit: Summary report*. TM-2010-216114. http://ston.jsc.nasa.gov/collections/TRS/_techrep/TM-2010-216114.pdf (accessed October 22, 2013).

Whitmire, A. M., L. B. Leveton, L. Barger, G. Brainard, D. F. Dinges, E. Klerman, and C. Shea. 2009. *Evidence report: Risk of performance errors due to sleep loss, circadian desynchronization, fatigue, and work overload*. http://humanresearchroadmap.nasa.gov/Evidence/reports/Sleep.pdf (accessed January 2, 2014).

Williams, D.R. 2011. *The Apollo 1 tragedy*. http://nssdc.gsfc.nasa.gov/planetary/lunar/apollo1info.html (accessed January 8, 2014).

Wu, H., J. L. Huff, R. Casey, M.-H. Kim, and F. A. Cucinotta. 2009. *Risk of acute radiation syndromes due to solar particle events*. http://humanresearchroadmap.nasa.gov/Evidence/reports/ARS.pdf (accessed January 8, 2014).

4

Risk Acceptance and Responsibilities in Human Spaceflight and Terrestrial Activities

Society demonstrates a willingness to tolerate high levels of risk to individuals who participate in certain types of activities. Every day, firefighters, law enforcement officers, other first responders, and military service members put their lives and health at risk in defense of persons, property, national security, or other compelling public interests. Other individuals volunteer to participate in biomedical research, which can pose significant health and safety risks. Additionally, many individuals engage in other high-risk occupational and recreational activities with limited external oversight.

As described in Chapter 1, the National Aeronautics and Space Administration (NASA) asked the Institute of Medicine (IOM) to convene the Committee on Ethics Principles and Guidelines for Health Standards for Long Duration and Exploration Spaceflights to identify ethics principles and develop an ethics and policy framework to guide decisions about long duration and exploration space missions, when risks associated with working conditions fail to meet current health standards or when uncertainty prevents development of adequate health standards. As part of its charge, the committee was asked to identify "models or examples of other situations with unknown health risks (or risks that could exceed current standards) that could inform NASA policy and, if so, how?"

In identifying other occupations and situations from which to draw, the committee considered the multiple roles of astronauts and NASA. As key figures in high-risk missions, astronauts concurrently serve a multitude of roles and responsibilities. Astronauts are employed as part of a mission crew, operating as a team to minimize risks to individuals and improve the likelihood of mission success. Each NASA crew performs a variety of mission functions, which may include flight management, operations, repairs, retrievals, and research (such as medical experiments). Beyond mission-related responsibilities, astronauts serve in a number of public roles, including federally sponsored explorers. Given that space missions are pursued for public good by a federal agency, astronauts are also public servants. Astronauts often participate in research as investigators and also as research volunteers. Similarly, NASA, as a federal agency, plays numerous roles (e.g., employer, research sponsor, international partner, and science educator) and bears responsibilities to various stakeholders with interests in space exploration.

The presence of highly uncertain and unquantifiable health and safety risks is not unique to human spaceflight. Many domains have negotiated similar challenges in decision making about risk acceptability. However, the degree to which ethics principles are incorporated into a formal decision record varies greatly by occupational domain and profession within a domain. The committee consulted the scientific literature about existing frameworks and spoke with many experts in other high-risk occupations, but was not able to identify existing ethics frameworks or decision-making models for specific occupations that were directly and wholly applicable to decisions about health standards for long duration and exploration spaceflight.

From existing ethics frameworks and available occupational health standards, the committee sought to identify common factors that could be used to inform committee deliberations about ethics principles and frameworks. The committee found examples in occupational health, research, and exploration that provided useful points for comparison. The first three sections of this chapter examine these domains for useful examples of risk management strategies and duties and, to some extent, identify relevant societal interests. The last section draws from these examples to categorize and synthesize examples of common factors that appear to influence decisions about risk management in terrestrial settings.

This chapter does not provide an exhaustive list of occupations and activities that involve high risk to participating individuals. Rather, it provides a few key examples that were selected based on characteristics that are applicable to human spaceflight and because they provide useful insights about the types of factors that may affect decisions related to risk management. Additionally, the factors identified are meant to be illustrative and not an exhaustive list of all factors that could be relevant to discussions about the risks of human spaceflight.

RISK MANAGEMENT IN THE WORKPLACE

Efforts to protect the welfare of workers in the United States have been documented since the late 19th century and have led to significant improvements in worker health protections. As described by some in the literature, occupational health regulations and standards "aim to promote and maintain the highest degree of physical[,] mental and social well-being of workers in all occupations; to prevent decline in health caused by their working conditions; to protect workers in their employment from risks resulting from factors adverse to health; and to place and maintain workers in an occupational environmental adapted to their physiological and psychological capabilities" (Serra et al., 2007, p. 304). In many high-risk occupations, health regulations and standards, in addition to serving the interests of individual workers, may also protect the health and safety of others (including bystanders and co-workers), equipment and property, and the integrity of the work enterprise.

Depending on the specific job hazards, employers (including NASA) use a multi-tiered approach to comply with established exposure limits and health and safety regulations. Employees also may be obligated to comply with these regulations and limits, without an option to waive mandatory employee protections, using a variety of techniques and controls, such as:

- Elimination of the hazard or substitution of less hazardous materials in the work process;
- Engineering controls and redesign of the work environment to limit exposure (e.g., ventilation, enclosure and/or isolation of emission source, process control);
- Administrative controls (e.g., length of work in a particular area, decision protocols); and
- Personal protective equipment and adequate training on correct use (e.g., respirators and gloves worn by firefighters).

As an agency involved in high-risk activities, NASA is responsible for astronaut candidate selection and training, resource allocation, risk and research prioritization, determining mission feasibility (taking into account available resources and technological capabilities), and deciding whether associated risks to astronauts, crews, and programs are acceptable. Like other analogous high-risk professions, spaceflight involves substantial risk to the short- and long-term health of individuals during ground-based training and during missions, due to numerous uncertain or uncontrollable variables. The space environment includes unique and sometimes unpredictable hazards, including prolonged isolation, reliance on a closed environment, limited basic resources, and high levels of radiation (IOM, 2001). These conditions can have profound and, sometimes, lasting effects on an astronaut's physical, physiological, or psychological health, as described in Chapter 3. NASA actively engages astronauts during all phases of their training and work in issues regarding health and safety and provides regular updates related to risks associated with spaceflight (Behnken, 2013). Astronauts may choose not to participate in a specific mission, and this decision should have few, if any, repercussions for inclusion on future missions (Behnken, 2013).

Workplace Risk Management by OSHA and NASA

In 1970, the Occupational Safety and Health Act (OSH Act) was enacted to "ensure safe and healthful working conditions for every working man and woman in the nation insofar as practicable, [so] that no workers will suffer diminished health, functional capacity or life expectancy as a result of their work experience."[1] To implement this legislation, Congress mandated the creation of the Occupational Safety and Health Administration (OSHA) within the Department of Labor. OSHA promotes worker safety and health through multiple approaches, including the establishment and enforcement of health standards and regulations restricting workplace exposures and mandating exposure and health monitoring, training, and information dissemination. The OSH Act also created the National Institute for Occupational Safety and Health (NIOSH), which conducts research and makes practical recommendations to prevent worker injury, illness, and death (CDC, 2013).

OSHA regulations apply to many occupations including those in general industry, agriculture, maritime, and construction; however, OSHA regulations do not cover all workers. OSHA does not have oversight over self-employed persons, employees of state and local governments (unless covered through an OSHA-approved state plan, which operates under state jurisdiction), or other federal agencies that regulate worker safety under the authority of other federal laws (including work-

[1]The Occupational Safety and Health Act of 1970, P.L. 91-596 (December 29, 1970).

places in nuclear energy and weapons manufacture, mining, railroad, and aviation).

NASA is authorized by the National Aeronautics and Space Act of 1958 "to make, promulgate, issue, rescind, and amend rules and regulations governing the manner of its operations and the exercise of the powers vested in it by law"[2] related to the planning, direction, and conduct of aeronautical and space activities.[3] This effectively exempts NASA from OSHA regulations and authorizes the agency to use its own discretion for programmatic activities, including "health and medical policies and standards that 'regulate' aircrew and space flight crew selection, qualification, and health-related requirements in NASA research aircraft and spacecraft" (Williams, 2013). However, NASA follows OSHA regulations for its ground workforce (Williams, 2013). For example, NASA's Johnson Space Center has qualified for OSHA's Voluntary Protection Program Star Rating (NASA, 1999; DOL, 2014), reflecting compliance with OSHA standards and periodic audits. For spaceflight, NASA has developed its own safety and health standards, as discussed throughout this report (NASA, 2007, 2011). When not preparing for or actively participating in space missions, astronauts may be engaged in more traditional occupational roles at NASA such as program development, operations, and management, and OSHA regulations apply to them as to all other NASA employees.

It is important to note that—unlike OSHA and NIOSH where responsibilities for regulation and policy research are separate from enforcement activities—NASA is tasked with assessing, evaluating, developing, and enforcing standards and regulations related to the health of its active astronauts, among other obligations. As described in Chapter 1, NASA is tasked with studying the potential benefits and problems associated with aeronautical and space activities,[4] which include risks to, or impacts on, human health. In this context, NASA functions may be more similar to those of the military in that research, regulatory development, and enforcement are all within the purview of the military.

---

[2]51 U.S.C. 20113(a).
[3]The National Aeronautics and Space Act of 2010, P.L. 111-314 (December 18, 2010).
[4]The National Aeronautics and Space Act of 1958, P.L. 85-568 (July 29, 1958).

Health Standards

One mechanism by which to manage occupational risk includes health standards, which are subject to revision based on new scientific data and require ongoing assessment. As discussed in Chapter 2, health standards may be used to protect workers; guide design, research, and engineering activities; stimulate innovation; serve as criteria for job requirements; and provide a condition for collaborative efforts. For example, some standards are relevant to worker selection or assessment of ongoing fitness for the job, which may involve periodic health monitoring. Other health standards protect against exposure to harmful physical agents or work environments during the course of employment and are intended to provide both immediate protection against acute injury and lifetime protection against harm. Health standards may be mandatory (e.g., OSHA standards) or voluntary, although often accompanied by significant incentives for compliance (e.g., the National Fire Protection Association's [NFPA's] standards[5] or OSHA's meatpacking guidelines).

NASA's health standards were developed to promote "a healthy and safe environment for crewmembers, and to provide health and medical programs for crewmembers during all phases of space flight," and include fitness-for-duty standards, space permissible exposure limits, and permissible outcome limits (NASA, 2007, p. 8). Although NASA's health standards are unique to spaceflight, they are specific manifestations of workplace protections more generally provided for workers. Those protections reflect broader ethical and legal determinations concerning risk acceptance within the occupational domain. The following sections explore examples of fitness-for-duty standards and exposure limits in terrestrial settings.

Fitness-for-Duty Standards

Many high-risk occupations require both applicants and current members within a profession to meet specific fitness-for-duty standards, which typically assess "whether an individual is fit to perform his or her tasks without risk to self or others" (Serra et al., 2007, p. 304) and are based on the definition of essential job functions within a given profes-

---

[5]Some state OSHA rules may incorporate language from otherwise voluntary standards (e.g., NFPA standards), effectively mandating compliance (see, e.g., Alabama Municipal Insurance Corporation and Municipal Workers Compensation Fund, Inc., no date).

sion. For example, NFPA has set health standards that apply to both firefighter candidates as well as incumbents (i.e., current members within a profession) (NFPA, 2013a). Fitness-for-duty standards are also used by law enforcement (Quigley, 2008; Fischler et al., 2011), the military (NRC, 2006), and interstate truck drivers.[6] In addition to physical standards, fitness-for-duty evaluations may also include psychological evaluations and mental capacity screenings, especially in jobs with high psychological demands, such as law enforcement (Fischler, 2011) and the military (NRC, 2006), including submarine crews (Kennedy and Zillmer, 2012).

The stringency with which medical and physical ability standards apply may differ for job candidates versus active members of a profession. For example, NFPA draws a distinction between health standards for candidates and incumbents, stating that the "intent with incumbents with a medical condition is to rehabilitate them and only restrict them from performing those essential job tasks where their injury or illness would affect the safety of themselves or others on their crew" (NFPA, 2013a, p. 1). The Air Force has a similar approach to fitness-for-duty standards applicable to current fighter and test pilots. In essence, the Air Force's fitness-for-duty standards serve as a baseline for entry into a field with a waiver process to allow individuals to continue to work, if approved, with some medical conditions (U.S. Air Force, 2013).

Limiting Hazardous Exposures

A number of federal agencies set exposure standards or guidelines relevant to the hazards and occupations, including the Nuclear Regulatory Commission[7] and Federal Aviation Administration,[8] as well as NASA (NASA, 2007). OSHA sets workplace permissible exposure limits (PELs) for a variety of chemicals and other hazards. For example, OSHA's lead standard limits the conditions under which an individual is allowed to work when lead is present in the work environment. OSHA's PEL for lead is 50 micrograms per cubic meter of air as an 8-hour, time-weighted average.[9] If an employee's lead exposure exceeds the established PEL more than 30 days per year, the employer is required to im-

[6]49 C.F.R. 391.41.
[7]See, e.g., 10 C.F.R. 20, Subparts C and D.
[8]See, e.g., 14 C.F.R. 25.832.
[9]29 C.F.R. 1910.1025(c)(1).

plement engineering and work practice controls (including administrative controls) to meet the exposure level, factoring in feasibility.[10] Respirators must be used to supplement engineering and work-practice controls when these are insufficient to reduce exposure below the PEL.[11] The standard also requires employers to continually monitor employee exposures when baseline levels are exceeded and notify employees about excessive risk exposures and subsequent corrective actions.

Health standards may also require specific employer or employee action based on individual clinical observations if poor health outcomes may be linked to specific occupational exposures. For example, OSHA's hearing protection standard states that if required audiograms show a "standard threshold shift" in hearing potentially related to occupational noise exposure, the employer must provide hearing protectors even if the noise exposure is within required limits.[12]

Federal agencies may also have significant latitude in determining what health standards are necessary to address a specific risk or hazard. For example, OSHA's occupational and health standards must be "reasonably necessary or appropriate to provide safe or healthful employment and places of employment."[13] The U.S. Supreme Court has held that "reasonably necessary" requires OSHA to demonstrate a "significant" risk to employees that can be eliminated or lessened by a change in practices, with two important qualifications. First, OSHA is not required to establish a "significant risk" with scientific certainty.[14] Instead, the agency can regulate on the basis of the best available evidence.[15] Second, OSHA is responsible for determining what constitutes a "significant risk."[16] In its review of OSHA's benzene standard, the U.S. Supreme Court characterized broad dimensions of significant risk in terms of acceptable and unacceptable risk, stating that, for carcinogens, a reasonable person would find a fatality risk between 1/1,000 ("plainly unacceptable") and 1/1,000,000,000 ("plainly acceptable") over a working lifetime to be significant.[17] The wide range between "acceptable" and "unac-

[10]29 C.F.R. 1910.1024(e)(1)(i). In these cases, the employer must reduce exposures to the lowest feasible level and must comply with additional respiratory protection requirements.

[11]29 C.F.R. 1910.1025(f)(1)(ii).

[12]29 C.F.R. 1910.95(j)(3).

[13]29 U.S.C. 652(8).

[14]*Industrial Union Dep't v. American Petroleum Institute*, 448 U.S. 607, 656 (1980).

[15]29 U.S.C. 655(6)(b)(5).

[16]*Industrial Union Dep't v. American Petroleum Institute*, 448 U.S. 607, 655 (1980).

[17]29 U.S.C. 655(6)(b)(5).

ceptable" risk, as defined by the U.S. Supreme Court, leaves ample room for philosophical debate. Although health standards are generally set based upon reasonably mature scientific knowledge, establishing health standards may be challenging in the context of a high degree of risk uncertainty.

Monitoring and Surveillance Programs

The threat of harm from specific exposures may trigger specific monitoring and surveillance efforts. In some cases, the federal government has committed to long-term research to better understand the risks of certain workplace exposures and to provide long-term screening for possible adverse health effects. During the Deepwater Horizon oil spill in 2010, clean-up workers were exposed to various hazards including oil, particulates, and oil dispersants (IOM, 2010). The National Institute of Environmental Health Sciences initiated the Gulf Long-term Follow-up Study (GuLF STUDY) to collect and analyze data to determine both acute and long-term physical and mental health effects of nearly 33,000 participants related to the Deepwater Horizon oil spill (NIEHS, 2014a). The GuLF STUDY "collected questionnaire data about oil-spill clean-up related exposures, health at the time of the spill and at enrollment, and lifestyle and other factors that might confound associations between exposures and health" (NIEHS, 2014b). Biological samples were collected for future research; and clinical measurements, including measures of pulmonary function, were made. Some participants may submit to more comprehensive clinical exams, and cancer and mortality records are linked to state cancer registries and mortality records (NIEHS, 2014b).

Similarly, the Federal Coal Mine Health and Safety Act of 1969,[18] as amended by the Federal Mine Safety and Health Act of 1977,[19] required NIOSH to jointly administer a program for early detection and prevention of coal worker's pneumoconiosis with the Mine Safety and Health Administration (CDC, 2014). The resulting Coal Workers' Health Surveillance Program includes the Coal Workers' X-ray Surveillance Program, requiring operators of underground coal mines to provide chest x-rays to new miners "as part of a pre-placement physical examination or within

[18]The Federal Coal Mine Health and Safety Act of 1969. P.L. 91-173 (December 30, 1969).

[19]The Federal Mine Safety and Health Amendments Act of 1977. P.L. 95-164 (November 9, 1977).

six months after being hired" and then 3 years later to screen for pneumoconiosis (CDC, 2014). The program also requires operators to offer chest x-rays approximately every 5 years to all underground coal miners (CDC, 2014). If "black lung" disease is confirmed, then individuals in environments where dust concentration is more than 1.0 milligram per cubic meter of air may transfer to a mine where the dust concentration falls below this cutoff and have their exposures monitored frequently (CDC, 2014).

Other Duties and Responsibilities

In addition to health standards and surveillance or monitoring programs, other employer and societal responsibilities may apply. The U.S. workers' compensation statutes aim to assist individuals who are injured or sickened on the job. The general purpose of such funds are to replace lost wages, cover medical expenses associated with on-the-job injuries, and provide vocational rehabilitation if a worker cannot return to a previous position due to an occupational illness or injury (Guyton, 1999).

For some high-risk public service occupations, the risks taken on behalf of society provide justification for long-term commitments to promote and protect the health of individuals within specific high-risk professions. For example, long-term health care benefits are available for all individuals who have served in the active military (VA, 2014a). Through the Veterans Health Administration, the Department of Veterans Affairs serves more than 8.7 million veterans each year (VA, 2014b) and provides "inpatient hospital care, outpatient care, laboratory services, pharmaceutical dispensing, rehabilitation for a variety of disabilities and conditions, mental health counseling, and custodial care" (CBO, 2007, p. 1).

RISK MANAGEMENT IN RESEARCH WITH HUMAN PARTICIPANTS

After a series of highly visible abuses in human research during the past century, governments and society established regulations to protect individuals who participate in research. The U.S. Office for Human Research Protections defines "research" as "a systematic investigation, including research development, testing, and evaluation, designed to develop or contribute to generalizable knowledge."[20] Biomedical research is often the focus of scrutiny because it can pose significant risk to the health and well-being of research participants and "its findings can have important implications for health" (IOM, 2002a, p. 17). The moral and social purposes of research regulations and policies are to guide judgments about when risks to an individual are low enough to justify inclusion in research protocols that may cause harm to the individual for the benefit of society, especially if there is no direct benefit to the individual participant. Moreover, the regulations promote distribution of potential benefits to individuals who chose to participate in research (or at least to the groups from which participants are recruited) and encourage trust from policy makers and the public by ensuring ethical research practices.

Biomedical research that is conducted to further the role of human beings in space exploration has been part of the space program since the outset. Astronauts serve a wide array of research capacities, including investigator, research coordinator, study team member, and finally study participant. In some cases, astronauts serve as the principal investigator for the research but, in other cases, astronauts contribute as study team members or inflight research coordinators to carry out medical experiments and collect data at the direction of an on-ground principal investigator. Astronauts also have the option of serving as research study participants, allowing their biomedical information to be collected, de-identified, and analyzed for research purposes.

Within the context of research, NASA as a government agency also plays multiple roles. It funds and supplies resources (such as equipment and labs in space) to facilitate research, provides mandatory oversight to ensure compliance with all applicable regulations, and plays a role in researcher and participant selection. As described in Chapter 2, NASA's Human Research Program focuses on research in physiology, environment, and technology to better understand the risks and opportunities associated with human spaceflight (NASA, 2014).

---

[20]45 C.F.R. 46.102(d).

Regulations for Research Involving Human Participants

Although the committee was not tasked with establishing an ethics framework to govern astronauts as research participants in space,[21] federal regulations governing human participation in research provide some of the best understood and accepted examples of ethics principles that govern risk exposures to individuals. Building upon key documents describing ethical research conduct,[22] the U.S. Department of Health and Human Services has adopted extensive regulations that govern federally supported or conducted research involving human participants.[23]

Independent oversight is an essential component of ethical research involving human participants. Under the Common Rule,[24] any research protocol involving humans must be submitted to an institutional review board (IRB), which is an administrative committee formally designated to review, approve, and potentially modify research involving humans (HHS, 1993). IRBs serve as the principal representatives of the interests of potential research participants (IOM, 2002b), functioning independently of, but in coordination with, other research-related entities (HHS, 1993). IRBs oversee not only the initial approval of a research protocol, but are also responsible for continuing review of ongoing research.[25]

As a general rule in research, as risks to individual research participants increase, so do the necessary levels of review, oversight, and the requisite, potential benefits. The Common Rule states that research may be approved only if risks to participants are minimized, if the risks of participation in a specific research protocol are reasonable in relation to the anticipated benefits and "the importance of the knowledge that may reasonably be expected to result."[26] Additionally, IRBs may only approve research protocols that include equitable selection of participants, safeguard and document informed consent and voluntariness of participation, provide adequate provisions for monitoring data collection to ensure

---

[21]In general, NASA medical data are not considered research data and may be used to update existing health standards (Williams, 2013).

[22]For example, *The Belmont Report* (DHEW, 1979), *The Nuremberg Code* (Nuremburg Military Tribunal, 1949), federal policy for the Protection of Human Subjects (45 C.F.R. 46.101-124), and *The Declaration of Helsinki* (WMA, 1964).

[23]45 C.F.R. 46.101-409.

[24]"The Common Rule" applies to all federally funded research involving human research participants and provides general guidance about levels of acceptable risk in light of potential benefits, participant selection, and informed consent (45 C.F.R. 46.101-124).

[25]45 C.F.R. 46.109.

[26]45 C.F.R. 46.111(a)(1) and (2).

participant safety, and ensure adequate protections of participant privacy and data confidentiality.[27]

Specific types of high-risk research may require continuous monitoring and review of study data to protect participant health and safety. For example, the National Institutes of Health (NIH) requires Data and Safety Monitoring Boards (DSMBs) for multisite trials that involve potentially risky interactions (NIH, 1998). Within NIH, the primary responsibilities of a DSMB include evaluation of monitoring systems for data to maintain participant safety; ensure data validity and integrity; evaluate study progress; and issue recommendations about the continuation or termination of a trial (NIH, 1998). DSMBs may terminate studies early if results indicate a clear risk or benefit to the study participants (NIDCR, 2014).

Even if an IRB and institution approve the levels of risk included in a specific research protocol, potential human research participants do not need to accept those risks. With few exceptions, "no investigator may involve a human being as a subject in research … unless the investigator has obtained the legally effective informed consent of the subject or the subject's legally authorized representative."[28] Proper informed consent involves sufficient opportunity to consider the risks and benefits of the research and should "minimize the possibility of coercion or undue influence."[29] Among other requirements, basic elements of informed consent include a written explanation of the purpose, procedures, and expected risks and benefits of the research, as well as a description of how confidential records will be protected. For research involving more than minimal risk, researchers are required to explain whether any compensation or medical follow-up are available if any harm occurs.[30]

RISK MANAGEMENT IN UNFAMILIAR ENVIRONMENTS

As described briefly in Chapter 1, exploration often involves the quest for information or resources in unfamiliar environments, and can lead to innovations and discoveries in many fields. Human presence in unfamiliar environments is often accompanied by risks. For purposes of this study, one of the more relevant examples of exploration in unfamil-

---

[27]45 C.F.R. 46.111.

[28]45 C.F.R. 46.116. IRBs may waive the informed consent requirement under certain conditions, per 45 C.F.R. 46.117(c).

[29]45 C.F.R. 46.116.

[30]45 C.F.R. 46.116(a)(6).

iar environments includes deep sea diving, which includes commercial, scientific, and recreational operations. Like astronauts, divers are exposed to a variety of health hazards, such as asphyxiation, respiratory and circulatory problems, hypothermia, and physical injury. Some of the more serious risks to human health arise from physiological changes within the body when exposed to high or changing ambient pressure (OSHA, 2014b). For example, the partial pressure of nitrogen in tissues can lead to nitrogen narcosis, affecting a diver's judgment while submerged (OSHA, 2014a). Analogous to space exploration, risks associated with deep sea diving are affected by the number and length of missions (i.e., dives), environmental conditions (e.g., temperature and visibility of the water), and the nature of the tasks involved.

Different regulations apply to different classes of divers. OSHA regulations apply to commercial diving and focus on personnel requirements, general operation procedures, specific operations procedures, equipment procedures and requirements, and recordkeeping requirements.[31] Although the regulations themselves offer some evidence of prevailing social norms that influence health standard applicability, four exceptions to OSHA's commercial diving standards may also be instructive. First, the purpose of specific diving activities is important. OSHA's commercial diving standards do not apply to operations "performed solely for instructional purposes[,]"[32] if the diver does not exceed certain depth thresholds and uses specific equipment.[33] In justifying its rationale, OSHA explained that, unlike commercial divers, scuba diving instructors can choose their diving locations and environmental conditions (OSHA, 1977). The standards also include an exception for scientific diving (OSHA, 1982),[34] which is based on a similar rationale, but further cites voluntary compliance with "well-established, consensus standards of safe practice" within the educational/scientific diving community (OSHA, 1982, p. 53357; see also, Lang, 2013).[35]

Second, commercial diving regulations do not apply to operations "performed solely for search, rescue, or related public safety purposes by

---

[31]29 C.F.R. 1910.410-440.

[32]29 1910.401(a)(2)(i).

[33]Certain exclusions to the instructor exceptions apply when dives exceed specific diving ranges, include a particular diving mode, or specific equipment (OSHA, 1977).

[34]29 C.F.R. 1910.401(a)(2)(iv).

[35]For examples of consensus standards for scientific diving, see AAUS, 2013 and University National Oceanographic Laboratory System, 2009.

or under the control of a governmental agency."[36] This language reflects OSHA's determination "that safety and health regulation of the police and related functions are best carried out by the individual States or their political subdivisions" (OSHA, 1977, p. 37655).

Third, diving operations governed by federal human research protections are also excluded.[37] OSHA reasoned that human research participation is already subject to federal oversight designed to promote safety and health (OSHA, 1977). Moreover, OSHA's final rule argues that the long-term safety and health of divers are best served by continued scientific research and continuous learning and improvement, which are designed "to extend the safe limits of diving physiology and technology" (OSHA, 1977, p. 37655).

Finally, OSHA includes an emergency exception to the commercial diving standards. Employers are allowed to deviate from the regulations if necessary "to prevent or minimize a situation which is likely to cause death, serious physical harm, or major environmental damage," provided that certain administrative actions are taken.[38] Purely economic or property damages do not trigger the emergency provision (OSHA, 1977).

As discussed above, oversight for deep sea diving activities does not fall to the jurisdiction or financing of a single entity or agency. Although space exploration is becoming more of a commercial and international endeavor, a significant portion of the responsibilities, public investment, and varying levels of societal engagement and investment rests with NASA, as do the burdens of risk management decisions.

RISK MANAGEMENT IN HIGH-RISK ACTIVITIES

In many high-risk activities, the desire to protect something of perceived value may increase the threshold of acceptable risk in terrestrial settings. For example, society has justified high risks to the health and safety of firefighters and police officers during emergencies based on the desire to save lives and property, among other interests (see, for example, U.S. Army Combined Arms Center, 2010; Arizona Division of Emergency Management, 2012). In the military, operational objectives, such as special operations involving counterterrorism, may also justify

---

[36] 29 C.F.R. 1910.401(a)(2)(ii).
[37] 29 C.F.R. 1910.401(a)(2)(iii).
[38] 29 CFR 1910.401(b).

high-risk activities (DoD, 2011). Similarly, the Federal Emergency Management Agency (FEMA) regulations allow a limited number of emergency workers to be exposed up to five times the amount of radiation expected during "all occupational exposures," in order to avert the exposure of large populations to dangerous levels of radiation (FEMA, 2008, p. 54037). To prevent the development of catastrophic conditions that could significantly affect large numbers of people, the International Atomic Energy Agency (IAEA) has proposed a radiation exposure limit for emergency workers of 500 millisieverts (mSv), substantially larger than the 1-20 mSv range applied to routine operations and significantly larger than the 20-100 mSv limit for exposures to accidents that are not potentially catastrophic; even the 500 mSv value can be exceeded in the course of life saving actions, if the "expected benefits to others clearly outweigh the emergency worker's own health risks" (IAEA, 2011, p. 93). Similarly, the U.S. Environmental Protection Agency (EPA) allows for increased radiation exposure to workers protecting large populations or property vital to the public welfare, provided that the exposure is unavoidable and appropriate actions have been taken to reduce the dose and monitor the effects of exposure (EPA, 2013).

The need for a well-trained workforce has also been used in terrestrial settings to justify increased health and safety risks in specific contexts. Realistic and often dangerous training may be used to prepare U.S. service members for operations (U.S. Coast Guard, 2013), although the military's tolerance for risk during peacetime training missions is quite low. For example, the armed services do tolerate higher levels of risk in programs like Top Gun or Red Flag, which prepare individuals for combat deployments (U.S. Air Force, 2012). The proven benefits of realistic, high-risk training have led to its subsequent adoption by fire and police departments, including Live Fire Training (NFPA, 2013b; Feyst, 2014).

Finally, the need for technological and scientific advancement may also affect tolerated levels of risk. For example, many operational test flights, which exist to maintain the United States' technological advantage (Allenby and Mattick, 2009), involve novel aviation feats, which increase the number and potential severity of known and unknown risks.

In many cases where federal regulations allow activities associated with high-risk exposures, such regulations also include a mandatory process for notifying individuals about those risks. In responses to radiological dispersal devices and improvised nuclear devices, FEMA requires that responders be "fully informed" of exposure risks that may occur and that risk acceptance be voluntary (FEMA, 2008, p. 45037). For doses exceeding 50 REM, responders must be "fully aware" of acute and chronic cancer risks related to the exposure (FEMA, 2008, p. 45037). Similarly, the IAEA safety standards require that workers who volunteer for increased radiation exposures understand and accept the health risks associated with the exposure (IAEA, 2011). In the military, although pilots are expected to comply with orders to fly, the military might ask for volunteers on extremely dangerous missions, where the risk of fatality is very high (Cockerham and Cohen, 1981).

FACTORS AFFECTING DECISION MAKING ABOUT ACCEPTABLE LEVELS OF RISK

Decisions about what constitutes acceptable risk are highly contextual, and different people may interpret the same set of facts about risk differently. In its report *Improving Risk Communication*, the National Research Council (NRC) (1989) identified many factors that affect risk perception including familiarity with, and understanding of the risk; perceived control over a situation; severity and immediacy of the consequences; level of potential benefit; who the risk affects, including specific populations; dread; equity; media attention; and trust in institutions. All of these factors can influence the weighing of perceived or potential risk and benefit levels and, therefore, risk acceptance within the context of a specific activity.

The scope of socially acceptable risk and, therefore, permissible activities, reflects social values and norms for conduct that help distinguish between acceptable and unacceptable behavior. These values and norms seldom articulate specific ethics principles underlying societal preferences or limitations. Instead, they generally reflect a combination of nonnormative variables like historical accident or political maneuvering, as well as normative factors like tacit moral judgments that may reflect common ethics principles.

Governmental regulations and oversight relevant to risk management decisions generally focus on fairly and justly protecting individuals from certain risks, especially when those risks are taken in exchange for some level of societal benefit. When society's interests in protecting individuals from harm do not restrict individuals' and institutions' decisions to engage in specific high-risk activities, other responsibilities may emerge. When taken piecemeal, the previous examples may seem like unrelated illustrations that offer no consistent patterns or guidance from which to draw conclusions. However, when taken as a whole, the examples include some common ethical norms embedded within decisions about risk in terrestrial settings, which parallel many of the factors described by the NRC.

Factors Influencing Risk Assessment and Tolerance

No single, simple summary statement adequately describes our national willingness to accept risks or our commitment to reduce them. However, the committee identified a number of common factors that underlie decisions about risk across a wide range of oversight. This section examines some of these factors, including types and severity of risk; the presence of actual harm; types and distribution of potential risks and benefits; activity purpose; the nature of the relationship between individuals approving, and those subject to, risk; the presence of independent oversight; and feasibility. This list of factors is merely illustrative and is not meant to represent an exhaustive list of factors that may influence risk assessment and tolerance.

Type and Severity of Risk

Societal tolerance for risk of harm to an individual typically reflects moral judgments about the probability of an adverse event occurring and the severity of a specific harm or resulting loss. For example, the OSH Act requires employers to provide, to the extent practicable, workplaces free of recognized hazards, assuring that employees do not suffer material impairment of health as a result of their lifetime of exposures to toxic agents or hazardous conditions at work. Permissible exposure limits are established to limit risk to individual workers, thereby maintaining the health of the working population and the functionality of society. Unknown environmental conditions, as described in the deep sea diving example, can increase risks to individual workers and participants, which may trigger increased governmental regulation of risk acceptance within those fields.

Presence of Actual Harm

When actual harms to individuals do occur in the context of high-risk activities, social morals may dictate follow-up actions. For example, workers' compensation programs are designed to protect workers from the negative impacts of actual injury or illness resulting from job-related activities. In rare cases, including exposures to hazards during the Deepwater Horizon oil spill cleanup, the threat of actual harm led to additional research and monitoring responsibilities on behalf of the government (NIEHS, 2014b).

Potential Risks and Benefits and Their Distribution

Specific classes of potential benefits trigger higher levels of societal risk acceptance. As described in the context of law enforcement and first responders, serious threats to life and, to a lesser extent, property will often justify increased risk exposures to individuals. Similarly, threats to national security also justify participation in activities that carry serious risks of mortality and morbidity. The need for a well-trained workforce and technological and scientific advancement are also commonly used to justify increased risk to individuals engaged in high-risk activities.

In addition to the class of benefit, the number of individuals who may receive a potential benefit (or avoidance of harm) also plays a role in risk acceptability. For example, the IAEA allows for greater radiation exposure for workers in emergency situations to prevent severe outcomes, including loss of life and catastrophic conditions with significant impacts on humans and the environment (IAEA, 2011). The protection of large populations also justifies increased radiation exposures under EPA regulations (EPA, 2013). The equitable distribution of risk and benefit also affects regulations governing clinical research involving humans.[39]

Activity Purpose

In general, society appears to be more willing to impose limits on risks within the same environment if the risks are related to conditions of employment. Consider the example of commercial deep sea diving. Although commercial diving operations may occur under conditions similar to deep sea diving for recreational or instructional purposes, governmental regulation is more robust in the context of commercial diving. Similarly, federally supported research involving humans immediately triggers review of risks to individual research participants. These decisions seem to embody the societal perception that workers and research participants deserve or require additional legal protections to regulate

[39] 45 C.F.R. 46.111.

how decisions about risk acceptance are made. Although similar to potential benefits, activity purpose is distinct. Potential benefits are typically balanced against potential risks or harms to determine risk acceptability; whereas, activity purpose may be used to broadly encompass an activity within the scope of governmental regulation.

Presence of Independent Oversight

Society may accept additional levels of risks to an individual's health and safety if objective, independent review bodies are in place to protect that individual's interests. As described above, in research involving human participants, IRBs provide independent oversight to ensure that all research protocols do not present excessive risks to individual participants, and that individuals voluntarily consent to participate in research. The Common Rule may allow a potentially high-risk research protocol if a DSMB exists to detect evidence of harms to patient health and safety early on and throughout the study. Similarly, an independent oversight body typically reviews and must approve operational test flights before an individual pilot can consent to high-risk activities.

Independent oversight may also take the form of either an external or internal advisory body, which is not responsible for ultimate decision making. For example, federal advisory committees "provide advice that is relevant, objective, and open to the public" across a broad range of issues and topics covered by federal policies and programs (GSA, 2013). The distinction between an independent body that serves as an ultimate decision maker versus a body that serves in an advisory role is critical, and the committee was cognizant of this distinction as it considered the ethics governing decisions about health standards for long duration and exploration spaceflights.

Nature of the Relationship Between Individuals

The nature of the relationship between individuals affects the creation and scope of the duties incumbent on professionals, which influence the responsibilities to mitigate risk of harm and the risk levels that society deems acceptable. U.S. statutory and common law creates affirmative duties of care to prevent unreasonable loss or harm to others through an overt act or omission. The duty of care, which often includes actions to mitigate or limit risk, may arise from "special relationships," including relationships between employers and employees, ones who assume control or custody of another,[40] or if a fiduciary duty exists (Easterbrook and Fischel, 1993). Other examples of special relationships include the doctor-patient relationship (ACP, 2012) and attorney-client relationship (Small, 2009). In occupational health, concerns about coercion may stem from real and perceived power imbalances between employers and employees (Hogbin, 2006). Similar concerns about power imbalance, as well as information assymmetries and patient confidentiality, exist between doctors and patients (ACP, 2012). Individual employees may be willing to tolerate unsafe or hazardous conditions due to limited alternative employment options, concerns about job security, incomplete information about relevant risks, and lack of control over the working environment. Conversely, employers select their employees, set wages, create and oversee working conditions, assign work and determine work pace, evaluate employees, and initiate employee termination. Similar concerns about coercion and undue influence exist in research involving human participants.

Feasibility

Decisions about acceptable levels of risk are often impacted by considerations of the feasibility of reducing hazardous exposures. In some cases, feasibility limits are employed in an effort to drive acceptable risk to the lowest level possible. For example under several environmental health statutes the paradigm is to set the allowable limits as low as feasible, even if the risks become vanishingly small. The OSH Act's use of feasibility allows higher risk than more conservative applications of feasibility. The OSH Act authorizes standards for exposure to workplace hazards that will ensure "no employee will suffer material impairment of health or functional capacity" but only "to the extent feasible."[41] Federal courts have interpreted "feasibility" to include both technical and economic feasibility in the context of OSHA rulemaking, but OSHA is not required to balance the cost and benefits of a proposed rule.[42] Other regulatory agencies may select exposure levels that are "as low as reasonably achievable," for example, to ensure that specific exposures are "as far below the dose limits as practical, consistent with the purpose for which the licensed activity is undertaken, taking into account the state of technol-

---

[40]Restatement (Second) of Torts §§ 314A, 314B, and 320 (1965).
[41]29 U.S.C. 655(6)(b)(5).
[42]*ATMI v. Donovan*, 452 U.S. 490 (1981).

ogy, the economics of improvements in relation to state of technology, the economics of improvements in relation to benefits to the public health and safety, and other societal and socioeconomic considerations."[43]

SUMMARY

In its task to look for analogs and models of how other occupations and efforts deal with uncertain health risks and risk management decisions, the committee found a lack of explicit frameworks but a variety of examples from terrestrial settings that inform deliberations about health standards for long duration and exploration spaceflights. Many of the efforts are focused on avoiding harm to workers and others who are willing to take risks to protect society. Both the context of the risk and the extent to which government, employers, or others are involved also affect the nature of the responsibilities and risk management approaches. Moreover, these factors implicitly embody ethics principles relevant to NASA decision making and provide valuable insights about considerations when designing an ethics framework to guide decision making about health standards for long duration and exploration spaceflights.

Finally, it is important to emphasize that space travel—with almost predictable certainty—will be fatal in some cases. Even so, the United States has continued to engage in human space travel for more than half a century. As described in Chapter 1, it is not within the charge of this committee to opine on the ultimate justification for human spaceflight. However the committee recognized that astronauts have volunteered, and are likely to continue to volunteer, for missions despite the uncertainty concerning the risks they will face. The committee kept this in mind as it considered ethics principles, responsibilities, and decision frameworks that are relevant to decisions about health standards for human spaceflight in Chapters 5 and 6.

REFERENCES

ACP (American College of Physicians). 2012. The physician and the patient. In *American College of Physicians Ethics Manual*. Philadelphia, PA: ACP.

---

[43] 10 C.F.R. 20.1003.

Alabama Municipal Insurance Corporation and Municipal Workers Compensation Fund, Inc. No date. *NFPA Standards for Fire Services*. Montgomery, AL: AMIC.

Allenby, B., and C. Mattick. 2009. *Macroethical and social issues in emerging technologies and the military.* Presented at IEEE International Symposium on Sustainable Systems and Technology, Tempe, AZ. http://ieeexplore.ieee.org/xpls/abs_all.jsp?arnumber=5156788&tag=1 (accessed February 5, 2014).

AAUS (American Academy of Underwater Sciences). 2013. *Standards for scientific diving*. Dauphain Island, AL: AAUS.

Arizona Division of Emergency Management. 2012. *Emergency response and recover plan: Basic plan*. http://www.dem.azdema.gov/preparedness/docs/basic_plan.pdf (accessed March 24, 2014).

Behnken, R., M. Barratt, S. Walker, and P. Whitson. 2013. *Presentation to the Institute of Medicine, Ethics Principles and Guidelines for Health Standards for Long Duration and Exploration Spaceflights: Astronaut office*. PowerPoint presented at the second meeting of the Committee on Ethics Principles and Guidelines for Health Standards for Long Duration and Exploration Spaceflights. Washington, DC, July 25. http://www.iom.edu/~/media/Files/Activity%20Files/Research/HealthStandardsSpaceflight/presentationfinal2.pdf (accessed November 8, 2013).

CBO (Congressional Budget Office). 2007. *The healthcare system for veterans: An interim report*. Washington, DC: CBO.

CDC (Centers for Disease Control and Prevention). 2013. *The National Institute for Occupational Safety and Health* (NIOSH). http://www.cdc.gov/niosh/about.html (accessed February 19, 2014).

CDC. 2014. *Occupational respiratory disease surveillance: Coal Workers' Health Surveillance Program (CWHSP) resource list*. http://www.cdc.gov/niosh/topics/surveillance/ords/cwhsp-resources.html (accessed January 15, 2014).

Cockerham, W., and L. Cohen. 1981. Volunteering for foreign combat missions: An attitudinal study of U.S. Army paratroopers. *The Pacific Sociological Review* 24(3):329-354.

DoD (Department of Defense). 2011. Special Operations. Joint Publication 3-05. Washington, DC: Joint Chiefs of Staff. http://www.dtic.mil/doctrine/new_pubs/jp3_05.pdf (accessed March 24, 2014).

DOL (Department of Labor). 2014. *OSHA voluntary protection programs: Current federal and state-plan sites*. https://www.osha.gov/dcsp/vpp/sitebystate.html (accessed January 15, 2014).

Easterbrook, F., and D. Fischel. 1993. Contract and fiduciary duty. *Journal of Law and Economics* 36(1):425-446.

EPA (Environmental Protection Agency). 2013. *PAG manual: Protective action guide and planning guidance for radiological incidences*. Washington, DC: EPA.

FEMA (Federal Emergency Management Agency). 2008. Planning guidance for protection and recovery following radiological dispersal device (RDD) and improvised nuclear device (IND) incidents. *Federal Register* 73(149): 45029.

Feyst, M. 2014. *Live-fire training: Much preparation for safety.* http://www.fireengineering.com/articles/2013/01/life-fire-training-much-preparation-for-safety.html (accessed January 15, 2014).

Fischler, G., H. McElroy, L. Miller, S. Saxe-Clifford, C. Stewart, and M. Zelig. 2011. The role of psychological fitness-for-duty evaluations in law enforcement. *The Police Chief—The Professional Voice of Law Enforcement* 8(LXXVIII).

GSA (General Services Administration). 2013. *The Federal Advisory Committee Act (FACA) brochure.* http://www.gsa.gov/portal/content/101010 (accessed February 27, 2014).

Guyton, G. 1999. A brief history of workers' compensation. *The Iowa Orthopaedic Journal* 19:106-110.

HEW (Department of Health, Education, and Welfare). 1979. *The Belmont Report: Ethical principles and guidelines for the protection of human subjects of research.* Washington, DC: DHEW. http://www.hhs.gov/ohrp/policy/belmont.html (accessed February 5, 2014).

HHS (Department of Health and Human Services). 1993. Chapter 1A: Jurisdiction of the Institutional Review Board. In *Institutional Review Board Guidebook.* Washington, DC: HHS.

Hogbin, G. 2006. *Power in employment relationships: Is there an imbalance?* Wellington, NZ: New Zealand Business Roundtable. http://archive.hrnicholls.com.au/articles/hrn-hogbin1.pdf (accessed February 5, 2014).

IAEA (International Atomic Energy Agency). 2011. Safety standards: Radiation protection and safety of radiation sources—international basic safety standards. In *General Safety Requirements Part 3.* Vienna, Austria: IAEA.

IOM (Institute of Medicine). 2001. *Safe passage: Astronaut care for exploration missions.* Washington, DC: National Academy Press.

IOM. 2002a. *Integrity in scientific research: Creating an environment that promotes responsible conduct.* Washington, DC: The National Academies Press.

IOM. 2002b. *Responsible research: A systems approach to protecting research participants.* Washington, DC: The National Academies Press.

IOM. 2010. *Assessing the effects of the Gulf of Mexico oil spill on human health: A summary of the June 2010 workshop.* Washington, DC: The National Academies Press.

Kennedy, C. H., and E. A. Zillmer. 2012. *Military psychology: Clinical and operational applications*, 2nd ed. New York: The Guilford Press.

Lang, M. A. 2013. *Diving in extreme environments: The scientific diving experience.* PowerPoint presented at the second meeting of the IOM Committee on Ethics Principles and Guidelines for Health Standards for Long Duration and Exploration Spaceflights. Washington, DC, July 25.

NASA (National Aeronautics and Space Administration). 1999. *Johnson Space Center: Johnson Space Center shines with OSHA star rating.* http://www.nasa.gov/centers/johnson/news/releases/1999_2001/j99-20.html (accessed January 15, 2014).

NASA. 2007. *NASA space flight human system standard. Volume 1: Crew health.* NASA-STD-3001. https://standards.nasa.gov/documents/detail/3315622 (accessed November 8, 2013).

NASA. 2011. *NASA space flight human system standard. Volume 2: Human factors, habitability, and environmental health.* NASA-STD-3001. https://standards.nasa.gov/documents/detail/3315785 (accessed December 4, 2013).

NASA. 2014. *Human Research Program: Human health and safety.* http://www.nasa.gov/exploration/humanresearch/areas_study/overview/index.html (accessed January 15, 2014).

NFPA (National Fire Protection Association). 2013a. *Standard 1582: Standard on comprehensive occupational medical program for fire departments.* Quincy, MA: NFPA.

NFPA. 2013b. *Standard 403: Live Fire Training Evolutions in Structures.* Quincy, MA: NFPA.

NIDCR (National Institute of Dental and Craniofacial Research). 2014. *Data and Safety Monitoring Board (DSMB) guidelines.* http://www.nidcr.nih.gov/Research/ToolsforResearchers/Toolkit/DSMBGuidelines.htm (accessed January 15, 2014).

NIEHS. 2014a. *About the study: Gulf long term follow-up study.* https://gulfstudy.nih.gov/en/about.html (accessed January 14, 2014).

NIEHS. 2014b. *The GULF STUDY.* http://www.niehs.nih.gov/research/atniehs/labs/epi/studies/gulfstudy (accessed January 14, 2014).

NIH (National Institutes of Health). 1998. *NIH policy for data and safety monitoring.* Bethesda, MD: NIH.

NRC (National Research Council). 1989. *Improving risk communication.* Washington, DC: National Academy Press.

NRC. 2006. *Assessing fitness for military enlistment: Physical, medical, and mental health standards.* Washington, DC: The National Academies Press.

Nuremburg Military Tribunal. 1949. *Trials of war criminals before the Nuremberg military tribunals under Control Council Law No. 10., Vol II.* Washington, DC: U.S. Government Printing Office.

OSHA (Occupational Safety and Health Administration). 1977. Occupational safety and health requirements: Commercial diving operations. *Federal Register* 42:37650.

OSHA. 1982. Educational/scientific diving. *Federal Register* 47:53357.

OSHA. 2014a. *Commercial diving laws*. https://www.osha.gov/SLTC/commercialdiving/laws.html#table_1 (accessed March 24, 2014).

OSHA. 2014b. *Commercial diving: Dysbarisms and other selected health effects*.https://www.google.com/search?q=hyperbaric+conditions&rls=com.microsoft:en-us:IE-SearchBox&ie=UTF-8&oe=UTF-8&sourceid=ie7&rlz=117NDKB117NDKB_enUS552 (accessed March 24, 2014).

Quigley, A. 2008. Fit for duty?: The need for physical fitness programs for law enforcement officers. *The Police Chief—The Professional Voice of Law Enforcement* 6(LXXV).

Serra, C., M. Rodriguez, G. Delclos, M. Plana, L. Gomez Lopez, and F. Benavides. 2007. Criteria and methods used for the assessment of fitness for work: A systematic review. *Occupational and Environmental Medicine* 64(5):304-312.

Small, D. 2009. Lawyer and witness: A special relationship. In: *Preparing Witnesses: A Practical Guide for Lawyers and Their Clients,* 3rd Edition, edited by D. I. Small. Chicago, IL: American Bar Association.

University National Oceanographic Laboratory System. 2009. Research vessel safety standards. UNOLS: Narragansett, IR. https://www.unols.org/publications/manuals/saf_stand/2009RVSS_web2012updates.pdf (accessed March 23, 2014).

U.S. Air Force. 2012. *Nellis Air Force Base: 414th combat training squadron "Red Flag."* http://www.nellis.af.mil/library/factsheets/factsheet.asp?id=19160 (accessed March 24, 2014).

U.S. Air Force. 2013. *Aerospace medicine: Medical examinations and standards*. Air Force instruction 48-123. http://static.e-publishing.af.mil/production/1/af_sg/publication/afi48-123/afi48-123.pdf (accessed March 23, 2014).

U.S. Army Combined Arms Center. 2010. *Disaster response staff officer's handbook:Observations, insights, and lessons*. Ft. Leavenworth, KS: Combined Arms Center. http://www.uscg.mil/hq/cg5/cg534/nsarc/Disaster%20Response%20Response%20Staff%20Officers%20Handbook%20(Nov%2010)8.pdf (accessed March 24,2014).

U.S. Coast Guard. 2013. *United States Coast Guard standard operating procedures for the Coast Guard Training System: Volume 14—High-risk training*. Washington, DC: U.S. Coast Guard. http://www.uscg.mil/forcecom/training/docs/HRT_SOP_01_Feb_13v3.pdf (accessed March 23, 2014).

VA (Department of Veterans Affairs). 2014a. *Health benefits: Veteran eligibility*. http://www.va.gov/healthbenefits/apply/veterans.asp (accessed January 15, 2014).

VA. 2014b. *Veterans Health Administration*. http://www.va.gov/health (accessed February 27, 2014).

Williams, R. 2013. *Charge to the IOM Committee on Ethics Principles and Guidelines for Health Standards for Long Duration and Exploration Spaceflights.* PowerPoint presented at the second meeting of the IOM Committee on Ethics Principles and Guidelines for Health Standards for Long Duration and Exploration Spaceflights. Washington, DC, May 30. http://www.iom.edu/~/media/Files/Activity%20Files/Research/HealthStandardsSpaceflight/Rich%20Williams.pdf (accessed November 8, 2013).

WMA (World Medical Association). 1964. *Declaration of Helsinki.* Ferney-Voltaire, France: WMA. http://www.wma.net/en/30publications/10policies/b3 (accessed February 5, 2014).

5

Recommendations for Ethics Principles

To begin developing an ethics framework for human activities involving risk and uncertainty, a series of questions needs to be asked to help define the objectives and parameters: What are the risks involved?; What is their likelihood and magnitude?; and Who bears them? What are the potential benefits of the targeted activities?; To whom will the benefits accrue?; and What values are to be safeguarded in the process? Who are the stakeholders involved?; To whom should the framework apply?; and Under what conditions?

As outlined in the 2001 Institute of Medicine report *Safe Passage*, the new challenges that will be faced in long duration and exploration spaceflight necessitate a relook at the ethics principles for these missions:

> Current ethical standards for clinical research and practice with astronauts were developed in an era of short space missions when repeat missions were the norm and a return to Earth within days was possible. In future missions beyond Earth orbit, however, a diverse group of astronauts will travel to unexplored destinations for prolonged periods of time. Contact with Earth will be delayed, and a rapid return will be impossible. Long-duration missions beyond Earth orbit, space colony habitation, or interplanetary travel will create special circumstances for which ethical standards developed for terrestrial medical care and research may be inadequate for astronauts. These ethical standards may require reevaluation. (IOM, 2001, p. 173)

Astronauts participating in long duration and exploration spaceflights will face a number of significant risks, including the impairment of health, as elaborated in previous chapters. Other individuals and groups are also potentially affected by the health consequences of these missions, including astronauts' families, astronauts hoping to participate in future missions, and the National Aeronautics and Space Administration (NASA) as the institution responsible for the human spaceflight program. NASA as a governmental entity has institutional responsibilities to manage risk effectively and derive maximum benefit from public expenditures. As part of that responsibility, NASA confronts the potential threats to ongoing and future missions that could result from emergent health problems in crew members. In turn, the halting or slowing of space exploration also affects the space program, its near-and longer-term goals, and those involved in the program, including astronauts preparing for future missions.

More broadly still, there are many important ways that the American public may be affected as stakeholders by the potential health consequences of long duration and exploration space missions. One related consideration concerns the ways that space exploration resonates with cultural and national values. Space missions, particularly those that are novel, are often the focus of national attention, and astronauts are perhaps the most visible embodiments of the cultural and national significance of space exploration. Successful missions, with the safe and healthy return of astronauts, represent great achievements and are a source of national pride. Despite risks to human life and health during exploration, many nations and their people have continued to explore. The growth of the commercial human space industry speaks to this imperative. Additionally, despite the well-known risks of spaceflight, including loss of life during both the Apollo and Space Shuttle programs, the recent astronaut applicant pool was one of the largest in history, with 8 astronauts selected out of more than 6,000 applications (NASA, 2013b). However, while there may not be known limits to the human imagination and initiative for exploration, the public still expects that NASA and the new commercial space companies will invest appropriately in protecting the welfare of crew and passengers. Harms to astronauts, whether as a result of launch catastrophes, errors, or unanticipated risks, could threaten the confidence that stakeholders place in the institutions charged with carrying out these missions and could jeopardize funding support for specific programs. That confidence is all the more important to safeguard where an institution's activities serve a unique social mission, and where there are no alternative institutions that can accomplish the same objectives and goals. Finally, the public's confidence extends beyond NASA as the primary institution involved in spaceflight to a more general trust in government.

NASA's health standards are based on current knowledge of the health risks and are updated as new knowledge becomes available. However, in contemplating long duration and exploration human spaceflight missions, NASA has to make decisions on how to handle missions in which those standards could not be met and astronauts could be exposed to higher levels of risk. Thus, NASA confronts three levels of decision making regarding health standards. In this decision framework (described more fully in Chapter 6), the first and broadest decision is *whether and under what conditions missions that are likely to involve greater risks to astronaut health and safety than health standards allow are ethically acceptable.* As described in previous chapters, long duration and exploration missions will entail significant risk and are almost certain not to meet some of NASA's health standards, as well as having the potential for confronting new and unknown risks for which future health standards will need to be set. In cases where health risks are recognized and health standards exist but proposed missions will likely not meet those limits, NASA must choose among three options: (1) apply the standards (thus foreclosing such missions given existing risk mitigation capabilities), (2) grant exceptions for individual missions, or (3) create new health standards that would apply only to long duration and exploration missions. The committee makes recommendations about the appropriate course of action among these options in Chapter 6.

Should NASA decide that long duration and exploration missions are ethically justifiable, the next level involves consideration of the *design of a specific mission in order to meet the ethics principles and obligations required to make it acceptable to fly.* Specific missions will carry unique potential benefits and opportunities, as well as their own risks and challenges. Evaluation of specific missions will consider factors such as characteristics of the destination, mission goals, duration (dependent on destination, propulsion systems, and system reliabilities), health risks, environmental risks, and feasibility of risk mitigation. Risks and risk mitigation strategies will be assessed for all elements of the mission.

Assuming a particular mission can be designed to meet the criteria of ethical acceptability, the third level of decision making concerns *the criteria and process for ethically acceptable selection of individual astronauts and composition of the crew.* Such recruitment requires an acknowledgment that individuals will vary in their susceptibilities to risk and the skills they bring to each mission.

This chapter provides a set of ethics principles that can guide the negotiation of each of the three levels of decision making. Each of the principles is relevant to decisions regarding the health standards—for example, whether a given risk level is acceptable in light of impact on astronaut health, available mitigation strategies, and anticipated benefits from the mission. Chapter 6 begins with a description of the processes that reflect the application of the ethical principles—for example, the process by which health standards are made and refined, as well as the process by which individual astronauts consider decisions about acceptable risk and their participation in missions. The development of the principles and processes for their application are designed to take into consideration the inherent uncertainty and unknowns in long duration and exploration spaceflight.

The sections below identify and discuss the principles of avoiding harm, providing benefit, determining favorable balance of risk and benefit, respect for autonomy, fairness, and fidelity. Each section begins with a short description of the principle, followed by some brief examples of how the principle is used and applied, and concludes with a focus on the principle in the context of long duration and exploration spaceflight. The following chapter examines the responsibilities needed for their implementation (informed decision making, continuous learning, independent analysis, transparency) and recommends a decision-framework for applying the ethics principles and responsibilities in setting and implementing health standards for long duration and exploration spaceflight.

OVERVIEW

There are numerous possible approaches to analyzing and addressing ethical issues, in whatever area of human endeavor they are faced. Among the challenges of reaching consensus around general approaches to addressing ethical issues is disagreement at the level of ethical theory. Over the course of centuries, philosophers have debated the relative merits and shortcomings of the major theories of Western moral and political philosophy: utilitarianism, duty-based approaches, virtue-based theories, and others. One successful approach to avoiding the need for a single ethical theory is to focus on mid-level principles rather than the theory to which they belong or from which they are derived. This approach has proven to be particularly successful when used by expert committees or commissions made up of individuals with diverse commitments to find common ground on how best to approach challenging ethical issues in the context of public policy. This was the approach and reasoning used by such landmark commissions as the National Commission for the Protection of Human Subjects in Biomedical and Behavioral research and the principle-based foundation it articulated in *The Belmont Report* (HEW, 1979). The committee adopted a principle-based approach in recognition of its accessibility and applicability for its task. Throughout the discussion of principles and their applications there are references to duties and responsibilities, and it is the committee's view that such duties are consistent not only with the principles identified but with a range of ethical theories as well, without any intended or unintended endorsement of a particular ethical theory.

The ethics principles described below draw on bioethics principles explored in the context of protection of individuals who participate in research. They have been refined over decades (HEW, 1979; Beauchamp and Childress, 2013). Their basis and the extent to which they represent a sufficient moral framework have been debated, but proponents argue that such principles are at least reflective of the so-called common morality (Gert, 1998; Beauchamp, 2003). The first three of these principles—avoiding harm, providing benefit, and favorable risk and benefit balance—are important both individually and as they relate to each other, as described below. The other three principles—respect for autonomy, fairness, and fidelity—each represent concepts that underpin an important aspect of the ethics of health standards in the context of long duration and exploration spaceflight. The principles were not created for this context nor are they unique to it, but rather each principle has been articulated and refined through a collection of scholarship and through the lessons learned in the application of these principles in a range of policy arenas including clinical medicine, biomedical research, laboratory science, public health, and health policy.

Much of the discussion of the ethics principles in this report draws on the approach and lessons gained in examining the ethics of research involving human participants, which shares at least some parallels with the issues raised in consideration of the health risks and mission benefits experienced in the course of human spaceflight. For example, in much of the research with human participants, the research participants agree to bear the health risks, some of which may be unknown or uncertain, for benefits that will be realized largely, if not exclusively, by society. Potential participants are asked to make a decision through the process of informed consent that includes disclosure of information and voluntary agreement to participate. Similarly, astronauts bear the risks, some of which are unknown or uncertain, for benefits that may be realized by the government and the general public; their participation is voluntary and based on shared decision making informed by disclosures and best available analyses.

The ethics principles identified as foundational to research with human participants are the legacy of a series of highly publicized occurrences of systematic violations of human participants in social science and medical research experiments. The disclosure of these exploitive uses of human participants led to the passage of the 1974 National Research Act (P.L. 93-348), which created the National Commission for the Protection of Human Subjects of Biomedical and Behavioral Research. One of the Commission's most important and influential publications, *The Belmont Report*, identified three fundamental ethical principles: respect for persons, beneficence and justice. It also identified processes for their applications: informed consent, assessment of risks and benefits, and selection of participants. These processes provide an analytical framework for conducting biomedical and behavioral research involving human participation (HEW, 1979).

Respect for persons reflects a belief that all individuals must be treated as autonomous agents, and those with reduced autonomy should be protected. Decisions to participate in research should be made voluntarily and with adequate information. Beneficence requires that research does not harm those who participate, maximizes potential benefits, and minimizes potential harm to participants. Investigators, therefore, have a responsibility to gather relevant information and conduct adequate assessment of risks and benefits, and to proceed only if there appears to be a "favorable ratio." Justice or "fairness in distribution" includes two prongs: (1) the selection of research participants be based on criteria specific to the research question rather than a function of availability, opportunity, or compromised position (e.g., socioeconomic vulnerability, poor health, age, or other disadvantage or incapacity); and (2) potential benefits of research be equitably distributed.

AVOIDING HARM

A fundamental principle in moral philosophy is nonmaleficence, which identifies ethical duties to avoid causing harm to others. Most analyses of the principle include additional ethical duties to remove existing harms and prevent harms that may be likely to occur. The principle of avoiding harm to others has a clear place in Western moral philosophy, whether from moral theories based on duties (Kant, 1998), utility (Mill, 1879), virtue (Aristotle, 2009), moral intuition, or more recent approaches based on a so-called duty of care in which it is argued that moral obligations derive from and relate to our relationships with others. In concrete terms, avoiding harm means that individuals have moral obligations (meaning they must be done as a matter of morality) to avoid causing harm to others, with harm generally defined as causing a setback to interests (Feinberg, 1986). Such setbacks to interest can include physical harm, psychological harm, harm to property, financial harm, and others. For example, all individuals have moral duties not to hit others with baseball bats, take their money, damage or steal their property, and so on—lessons we learn starting early in life.

Less obvious are the related moral duties to prevent or remove harm. Obligations to prevent harm are closely related but not entirely overlapping with duties of avoidance. Failing to erect a fence around a swimming pool in an area frequented by small children is by any account irresponsible, but it is not equivalent to pushing a small child into the pool. One is failing to prevent a possible harm from occurring, while the second would be failing to avoid causing harm; both are moral failings, but they are distinct, nonetheless. Removing harm has a different status yet again, and to continue the example, would entail obligations to rescue the struggling child who has fallen in the pool. Does a passerby have a moral duty to rescue? It is hoped that people will come to the aid of others but society expects it only in a limited range of circumstances such as when the risk of rescue is consistent with professional duties or the risk of the action required is far outweighed by its likely benefit. Removing harm may be considered as a part of the principle of beneficence (doing good for others) rather than as part of a principle to avoid harm.

The principle of avoiding harm inevitably leads to questions about the extent to which a person or entity must act to reduce risks or prevent harm to another in order to have fulfilled the obligation, particularly in the face of uncertain risks. The course of public policy making in the United States offers numerous examples related to reducing or limiting the risk of harm. In common law, the duty of care includes a responsibility to prevent unreasonable loss or harm to others through an overt act or omission. In *Industrial Union Dept., AFL-CIO v. American Petroleum Institute*, the U.S. Supreme Court ruled that the Occupational Safety and Health Administration's (OSHA's) benzene exposure limits had to be supported by substantial evidence that indicated "a significant risk" of harm was more likely than not, and that it was reasonable to "use conservative assumptions … risking error on the side of overprotection, rather than underprotection."[1]

In some cases, such as long duration and exploration spaceflights, there are many unknowns and a great degree of uncertainty regarding the nature and extent of the risks. For those types of missions, there will probably be both "known unknowns" (known risks of unknown extent, i.e., galactic cosmic radiation) and "unknown unknowns" (unanticipated risks). In such cases, the unknowns need to be addressed, to the extent feasible, in policy and program decisions using an iterative approach to worker protection (or protection of the general public) (see Box 5-1). There are no failsafe measures for neutralizing "unknown unknowns;" rather, the task is to minimize and mitigate them and to learn from prior flights to the extent possible.

BOX 5-1
Addressing Uncertainty in Avoiding Harm

Reasonably reliable, frequency-based risk probabilities used in risk assessment are often not available when activities involve novel technologies, exposures, or previously unrecognized harms. In such circumstances, policy makers often use cautious health risk estimates based on upper boundary estimates. A cautionary approach favors initial emphasis on protection with gradual risk escalation that is based on a commitment to continuous learning and using the experience and knowledge gained to inform future decisions.

This approach incorporates unquantifiable but credible prospects of catastrophic harm in decision making. Because unexpected, irreversible harms are the types of events that generate significant social controversy, precautionary decision making can protect both the welfare of those exposed to health risks as well as the integrity of institutions charged with implementing a risky activity. Second, cautionary approaches allow risky activities to be structured in a way that diminishes uncertainty over time. Further research and cautious forays into uncertain realms allow collection of evidence that can inform more quantitative risk-benefit assessments. Lastly, caution–if applied appropriately–distributes risk and benefit more fairly. Often, risks associated with exposure and costs associated with their remediation are borne by different individuals. Where risks associated with a health exposure are involuntarily endured, precaution provides a shield against unjust burden and provides the opportunity to consider the fair distribution of risks and benefits.

Drug developers and regulators negotiate uncertainties in two ways. First, they often employ "safety factors" when selecting initial doses to give to patients. For example, where novel drugs are given to patients for the

[1] 448 U.S. 607, 656 (1980).

first time in drug development, drug developers often introduce margins of safety into calculations of the starting dose to account for the possibility that toxicities may have gone undetected in prior animal studies. Second, they often stagger dosing between patients and across cohorts of patients to learn from the experiences and fully inform subsequent research. For example, the first patient's response to a drug might be monitored before a drug is given to a second patient; effects at low doses will be observed before groups of patients are given higher doses.

At NASA, as in many organizations where risks are uncertain, severe in magnitude, and irreversible, margins of safety are incorporated into risk estimates and are built into engineering and planning decisions. Additionally, research and continuous improvement cycles are implemented to ensure learning from experience and the escalation of risks as appropriate and ethically acceptable.

Putting the Principle in Context

In the context of professional or institutional rather than personal duties, obligations to avoid or prevent harm are equally important though often complicated to carry out. For example, when, if ever, is it acceptable for a physician to put a patient at risk for harm? Undergoing surgery for a life-threatening condition is one common example of such a tension. The usual analysis is that the benefit of saving the patient's life justifies the potential surgical risks, such as harms due to incisions and anesthesia—the patient both bears the risk and receives the benefit that comes with it. Even if the patient is fully voluntary in his or her agreement to undergo the surgery, the decision about the surgical technique and approach is made by the surgeon. That said, even such a clear-cut justification is subject to the patient's right to autonomous decision making, which includes accepting the potentially life-threatening consequences of foregoing surgery if it is the patient's wish to do so.

In other contexts, the harm that individuals are exposed to may be outweighed by the proposed benefits, but those potential benefits may accrue to others and not to those who are exposed to harm, as in early-phase drug development trials in humans. In such examples, there must be increased attention to obligations to avoid harm because there may not be any offsetting benefits to participating individuals and there may be circumstances that do not permit the research to go forward without further modification to risk. In settings where individuals are placed in harm's way for the benefit of others or society in general, policies of avoidance, protection, and removal of harm are often employed, whether by minimizing risk for research participants, requiring safety equipment and protective gear for firefighters, or limiting the type and duration of risky work for oil rig workers.

Focusing on the Principle in the Context of Spaceflight

Human spaceflight is undertaken with the knowledge that it exposes others (astronauts) to certain and uncertain risks of harm for benefits that will accrue largely, if not exclusively, to society as a whole, and so bears greater resemblance to research without potential benefits to those individuals who participate than it does to medical treatments that may bring risks along with desired benefits. This dynamic makes justification of the risks of human spaceflight more demanding than would be the case where risky undertakings have the potential to more directly benefit those assuming the risk. In the terrestrial context, where it may not be possible to completely avoid or prevent risk, there are often ways to remove harm or to rescue individuals who have come to harm. No such interventions are likely to be available during long duration or exploration class missions where there are many challenges for rescue and immediate return to Earth may be difficult or indeed impossible.

Any consideration of health standards for long duration or exploration types of missions must be guided by the principle of avoiding, preventing, or removing harm, which involves doing all that is feasible through vehicle design, safety processes, protective technologies, and other risk mitigation strategies. Furthermore, as described below, the balance of risks and benefits must be appropriate, their distribution fair, and the decision making to participate by astronauts fully informed. In addition to considerations of harms caused to the astronauts participating in spaceflight, the harms considered should also include the potential threats to spaceflight as an enterprise and to NASA as an institution. Avoiding these types of harm provides additional motivation for risk reduction because short- or longer-term harms to the health of astronauts, potential damage or loss of spacecraft, or failure of all or parts of missions could cause serious and significant setbacks to the agency and the future of space exploration.

PROVIDING BENEFIT

An ethics principle related to avoiding harm is the principle of providing benefit to others, often termed the principle of beneficence

(HEW, 1979; Beauchamp and Childress, 2013). It guides ethical behavior by promoting actions that provide benefit. This is especially important when there is risk of harm associated with the action being considered; the potential benefit is an important moral counterweight. The provision of benefit can be a clear obligation, as in the recognition of "good Samaritan" duties to come to the aid of others. In other contexts, doing good for others is commendable but not necessarily obligatory.

While there are clear distinctions between the moral status of a duty to refrain from knowingly causing harm and the articulation of a duty to provide benefit, they are not always easily separated. For example, sometimes the benefit can only be realized when accompanied by the risk of harm. Medical treatment is rife with such tensions (e.g., how much of a toxic chemotherapy agent can be tolerated by a patient with cancer; when is it acceptable to undertake risky surgery in the interest of treating illness, or in attempting to relieve suffering). As noted later in this chapter, the balance between risk and benefit is a key feature in the process of setting ethically appropriate health standards by NASA.

Putting the Principle in Context

The principle of beneficence has a longstanding place in the ethics of medicine and the physician-patient relationship. In much of medicine, the harms experienced in the course of treatment are expected to be offset by the benefits realized by the same individual. Not so in examples from other sectors, including many occupational settings where workers experience significant risk for benefits to be realized by employers, companies, stockholders, or society. Firefighters and police officers routinely put their lives and health at risk for the benefit of others and society. Military service members often face significant risks to their health and life for the benefits of national defense and protecting the nation's freedom and values, putting themselves at risk as a function of their professional duties and their obligations to each other and to their country. The benefits sought are often valuable and important, and their realization carries institutional responsibility—both for assuring that they are achieved and for minimizing the potential harms, costs, and negative consequences that come with them.

Focusing on the Principle in the Context of Spaceflight

Although individual astronauts have personal motivations and may receive benefits to participate in spaceflight, the benefits of human spaceflight, for the most part, accrue to society through technological and scientific advances, as well as to national and international pride and collaboration. NASA is founded on a mission to realize the benefits of space exploration, as noted in its vision statement, "To reach for new heights and reveal the unknown so that what we do and learn will benefit all humankind" (NASA, 2013a). In addition to the benefits realized by the space program, research and development sectors, and society, long duration and exploration missions may also yield information that will benefit the health of astronauts who participate in future missions.

Beneficence in the context of these missions entails planning, implementing, and following up on missions in ways that maximize the attainment of the multiple benefits. As noted below, satisfying the principle of beneficence cannot be considered in isolation, as only a function of the magnitude of the benefits obtained; rather, consideration must also be given to how those benefits are distributed. The principle of fairness clearly has a role to play, in that opportunities for participation in spaceflight and the distribution of the benefits achieved be distributed and shared equitably. NASA has put extensive effort into detailing, analyzing, and articulating the risks of long duration and exploration spaceflight; similar efforts are needed regarding the valuation of the benefits of these missions, as decisions related to the ethics of exceeding risks allowed by current health standards will need to weigh both.

FAVORABLE BALANCE OF RISK AND BENEFIT

There are many realms where individuals or groups are asked to take on health risks for the benefit of others, such as military service and emergency rescue work (see Chapter 4). Many areas of discovery and innovation involve risk for the benefit of others, including drug trials and test piloting. For these activities to be ethically justified, the value of what is achieved in their pursuit must morally redeem what potentially could be lost. This notion is often encapsulated in the principle of favorable balance of risk and benefit.

Individuals frequently make decisions about tradeoffs of risk and benefit as they contemplate whether to engage in behavior or activities that could result in harm to self. For some individual-centered decisions, there are few society-imposed restrictions for avoiding harm. However, there are many other circumstances where there are strong ethical imperatives to ensure that risk is favorably balanced with benefit. One of these is where individual risk taking imposes involuntary risk on others. Examples include laws about alcohol use and driving, restrictions on smoking in public places, environmental regulations, and occupational safety standards. Other areas where risky activities are undertaken in the name of some broader social objective include military service, rescue work, and participation in highly novel medical interventions (e.g., immunotherapies, cell-based interventions). Here, society (including members of the public, governmental institutions, and private-sector entities) has obligations to exercise good judgment in recruiting, training and informing, and engaging participants. Governments, in particular, are expected to ensure that the potential sacrifice of a sanctioned activity is adequately balanced by the benefits.

Space exploration entails risk to others, and it requires state approval of risk. It involves the former because, as noted previously, many other stakeholders are potentially affected by risks taken by individual astronauts, including fellow crew members, NASA, contractors, and any individuals and entities that benefit from the sustained enterprise of space exploration. Space exploration is state-sanctioned insofar as NASA is a public agency, and pursues exploration in the name of public interests.

Promoting a favorable balance of risk and benefit involves several activities:

- *Systematic assessment of risk and benefit:* This entails as thorough an accumulation of evidence as is feasible regarding risks and benefits, a systematic appraisal of that evidence, and a process for aggregating disparate forms of both quantitative and qualitative evidence. Risk assessment is greatly aided by deconstructing risk problems into their components and evaluating each. This includes systematically assessing long duration and space exploration risk components such as launch, microgravity exposures, radiation exposures, etc. Risk assessors must also bear in mind, however, that risk components can interact in ways that are additive or synergistic, and systematic risk assessment is only complete once cumulative risks are estimated.
- *Minimizing risk:* Standard setting and risk decision making should measure an activity's risk against feasible alternatives.

Activities are unacceptably risky when other safer means are available to accomplish the same objective. For example, a mission might be redesigned, more shielding countermeasures applied, crew members selected who are less susceptible to risks, and labor/work flows divided to minimize risk exposures.

- *Moral evaluation of residual risk:* There are simply no objective or value-free ways of deciding whether a given risk-benefit balance is favorable, whether in space exploration or in other risky domains such as medical research or military activities. Instead, risk-benefit determinations are a direct and explicit expression of the many values and beliefs that surround an activity. Qualities that should be inherent in risk-benefit decision-making processes and judgments are detailed below.
- *Monitoring and timely revision:* As any activity unfolds, events can alter a risk-benefit balance in unanticipated ways. For instance, a technical failure may impair the ability of a mission to accomplish its objectives, thus greatly diminishing the risk-benefit balance. A health event might occur (e.g., an injury) that greatly worsens risks for an astronaut. A new discovery may obviate the need for a continued mission. As much as possible, even for missions already under way, mechanisms should be designed so that events that might alter a risk-benefit balance are monitored to the extent possible, and missions revised accordingly.

Putting the Principle in Context

Decisions about balancing risk and benefit are informed by extensive research and management guidance developed through the fields of risk assessment and risk management. The steps of hazard identification, risk assessment, and decision making are continually refined, as new information becomes available.

For industries with known potential for exposures to hazardous materials, such as the chemical industries, nuclear energy, and drug development, quantifying and reducing the risks is paramount. Research, surveillance, and worker protection regulations have resulted in detailed risk profiles, permissible exposure limits, and preventive and mitigation measures. Policies and regulations aim to address worker protection in the event of even the most catastrophic conditions (e.g., nuclear power plant accidents).

For many occupations, such as rescue work, risks are harder to anticipate and quantify. Here, a crucial risk-mitigation strategy involves developing protocols and decision frameworks so that risk-benefit decidecisions can be informed by systematic moral and scientific thinking. Other realms encountering unquantifiable risks employ values, such as precaution, or decision rules, to promote non-arbitrary decision making.

Focusing on the Principle in the Context of Spaceflight

Health standards for long duration and exploration spaceflight, as well as decisions concerning specific missions, should demonstrate a favorable balance of risks and benefits. Balancing of risk and benefit could lead to establishing a high ceiling on acceptable risks for astronauts, especially when the endeavor involves pressing state interests and when astronauts have assumed risks voluntarily (see section on Informed Decision Making in Chapter 6). Yet, there are good grounds for establishing limits that also derive from the values and beliefs held by astronauts and members of the public. Risks to be considered should include those that affect the welfare of astronauts, as well as the enterprise of space exploration. Benefits to be considered should include those expected to accrue to society, as well as to future space travelers. Acceptable risk-benefit judgments should take into account both quantifiable risks and benefits, as well as those not quantifiable given current knowledge. Risk-benefit judgments should demonstrate the following qualities (Fischhoff et al., 1981):

- *Rationality:* Decisions should be grounded in evidence—where it is available—and logically joined with values and preferences.
- *Systematicity:* Decisions should be based on as thorough an accumulation of evidence as possible. Much of this evidence will be based on observational and experimental studies of risk, but evidence also can derive from other sources. For example, knowledge about the typical shortcuts users may use to apply a technology, or how risks will be experienced by different stakeholders, can help in estimating risk. In addition, it is important to have a good understanding of the values different stakeholders hold when the risks and benefits of an activity are being considered.
- *Explicitness:* Decision makers should strive to make clear the values and assumptions supporting their decisions, the evidence and reasoning used to reach them, and the uncertainties sur-

rounding such evidence. In the case of long duration and exploration space missions, decision makers need to render as explicit as possible not only the risk estimates, but also the nature and value of missions, the viability of alternative and safer strategies for achieving these ends, and why pursuit of those values is sufficient to redeem the health hazards associated with missions.

- *Coherence:* Risk acceptability judgments for long duration and exploration spaceflight should strive to be consistent with those in similar activity domains that are deemed to render morally and politically sound risk judgments. Deep sea diving and polar science missions, as discussed elsewhere in this report, provide useful examples.
- *Responsiveness:* Acceptable risk standards embed and express deep-seated values surrounding an activity, and acceptable risk determinations should strive to be responsive to the values of stakeholders who perceive their interests as implicated by the standards. A large body of evidence exists, for example, that publics are often less tolerant of risks that are associated with dread, uncontrollability, and surprise (Slovic, 1987; Sandman, 1989).

  In some respects, long duration and exploration space missions have properties that render the public more willing to countenance high risks; spaceflight is perceived as a socially valuable activity and astronauts assume risks voluntarily. In other ways, however, these space missions may provoke more restrictive attitudes toward risk. For example, the public tends to be more cautious about risks that are perceived as unjustly distributed. This caution might apply to spaceflight missions because the majority of mission planning is done by individuals who do not assume risks themselves. Furthermore, the public may also be more restrictive about risks that implicate important values. In many ways astronauts are living embodiments of national and cultural aspirations and values.
- *Political acceptability:* As noted above, the public's values may be in conflict or in flux. Advocates of occupational safety, for example, may have very different views of risk acceptability than astronauts themselves. Libertarians might champion the decision-making autonomy of astronauts, whereas paternalists may place an emphasis on the role of states to protect individuals from undue harm. Decisions and standard setting should strive to

consider, accommodate, and earn the confidence of credible skeptics who might question a particular decision, given their individual perspective.

RESPECT FOR AUTONOMY

The principle of respect for autonomy has been among the most fundamental principles of bioethics since the latter half of the 20th century (HEW, 1979; Faden and Beauchamp, 1986; Tauber, 2005; Beauchamp and Childress, 2013).[2] In addition to being a basis for much analysis related to the ethically appropriate treatment of individuals in health care settings and in biomedical research, it is the basis of the many public policies that seek to foster and protect individual self-determination in diverse contexts, including civil rights, consumer rights, contract and tort law, and criminal defense doctrines (Smith, 1982; Feinberg, 1986; Raz, 1986; Dworkin, 1988; Shapiro, 1988; Kahn, 1992; Rawls, 1993; Fallon, 1994).[3] This principle is rooted in respect for individual liberty and freedoms. These include the right to be left alone (a non-interference right), and most relevant for the discussion in this chapter, the right to make decisions for and about oneself without the unjustified interference of others.

Putting the Principle in Context

Physical Risks

There are many contexts in which individuals are given the liberty to make decisions, including those that may entail the risk of injury to themselves. On a personal level, these contexts include contact and extreme sports, thrill-seeking activities, cosmetic piercing, and tattooing. On a community level, individuals often volunteer to serve as firefighters, rescue workers, or other vocations in which they put themselves at

---

[2]*The Belmont Report* (HEW, 1979) states, "Respect for persons incorporates at least two ethical convictions: first, that individuals should be treated as autonomous agents, and second, that persons with diminished autonomy are entitled to protection."

[3]Fallon (1994) notes "[a]utonomy has been identified as a value underlying the constitutional protection of privacy, procedural due process, equal protection, and free exercise rights" and that "[a]utonomy also explains the antipaternalist sentiment generally dominating American law" (p. 876).

risk. Professionally, individuals join the military service or work in emergency situations or in jobs, such as mining, that involve higher than average risks. Many factors go into individual decisions about whether to accept higher degrees of risk including the availability and feasibility of alternative opportunities, the benefits and results of the endeavor, compensation, and safety plans.

In practice, law and ethics both recognize that there are situations where it is appropriate to restrict the liberty of autonomous individuals to consent to self-harming activities. In criminal law, a victim's consent to be injured may not always provide a defense for the person who inflicted the injury (Law Commission of England and Wales, 1995). In contract law, courts may refuse to enforce a contract that is so detrimental or injurious that it is "unconscionable," even if the contract was voluntarily entered (ALI, 1981). In tort law, courts recognize limits on individuals' ability to consent to an actual injury or to the risk of injury (ALI, 1979).

Yet even in contexts where individual autonomy is highly valued, such as health care and biomedical research, there are ethical and legal restrictions on individuals' powers of consent. Under federal regulations for the protection of human participants, individuals may only consent to research in which risks have been deemed appropriate, as determined by an Institutional Review Board (45 C.F.R. § 46.111(a)), thus limiting the individual's autonomy to consent to riskier research. Although patients have wide latitude to consent to elective surgeries, law and medical ethics impose limits (Bergelson, 2007). Striking the right balance has been the purview of much of the scholarship in research ethics and the promulgation of regulations for oversight of research involving human participants (Faden and Beauchamp, 1986; National Bioethics Advisory Commission, 1999; Mastroianni and Kahn, 2001; Moreno, 2001; 45 C.F.R. § 46 Subpt A), and courts struggle to enunciate principles to delineate appropriate boundaries of individual consent in many contexts.

Many regulations and procedures attempt to quantify risks, whether loosely through standards that simply require risks to be minimized or, more precisely, through numerical limits on permissible exposures. Although quantitative limits on the risks to which people can consent are often challenging to determine or implement, they are often a practical necessity. Even if people are willing to undertake risks, institutions still bear a responsibility to minimize risk and not invite people into activities that present undue risk.

Informational Risks

Autonomy is also expressed in the right of individuals to regulate the flow of information about themselves to others. The rights to privacy and confidentiality of health information derive from this principle (HEW, 1979; Faden and Beauchamp, 1986; Moskop et al., 2005; Beauchamp and Childress; 2013) and require mention because violation of these rights may not entail direct physical harms. A balance is often sought between individual autonomy and society's need for access to health information that can contribute to research, public health, and transparent, evidence-based policy formation and standard-setting (IOM, 2009). Because individuals may have important interests on both sides of the balance (e.g., benefitting from research), privacy and confidentiality[4] are not always accurately framed as being directly in conflict with data access and transparency. The challenges come in establishing limits while finding the balance in the flow of information so as to benefit individuals and society. The loss of privacy and confidentiality can cause psychosocial harm (e.g., embarrassment or stigmatization if sensitive medical information is released), potential economic harms (e.g., discrimination in employment or inability to purchase life insurance), or indirect physical harms (e.g., receiving suboptimal health care if concerns about privacy and confidentiality cause patients to withhold information from their health care providers).

Focusing on the Principle in the Context of Spaceflight

Considerations of the principle of autonomy are relevant to decisions about astronaut participation in a specific mission, operational planning, and the extent to which health information is collected and shared. Critical to an astronaut's decision is the right to be fully informed regarding health and safety risks of long duration or exploration space missions. Astronauts are highly trained and knowledgeable individuals who are

[4]Although privacy and confidentiality often are used interchangeably, they refer to distinct concepts. As stated by the Institute of Medicine report on the Health Insurance Portability and Accountability Act Privacy Rule, "[P]rivacy pertains to the collection, storage, and use of personal information and addresses the question of who has access to personal information and under what conditions" (IOM, 2009, pp. 16-17). "Confidentiality, though closely related to privacy, refers to the obligations of those who receive information in the context of an intimate relationship to respect the privacy interests of those to whom the data relate and to safeguard that information" (IOM, 2009, pp. 17-18).

provided in-depth briefings and opportunities to participate in NASA's risk management processes at many levels.

Preparing for future exploration missions will necessitate changes in the current modes and procedures for communications and interactions between crew members and ground control (NASA, 2010). Due to the nature and distance of future missions (e.g., missions to Mars), there are expected to be communication delays and associated technical difficulties that are not experienced on current missions. These communication delays may impact the quality of communications and team coordination in such a way as to require the crew to work semi-autonomously in order to maximize health and performance. Considerations will be needed regarding the extent to which individuals and teams have the discretion and freedom to conduct tasks, make decisions, solve problems, and carry out other general duties (Leach et al., 2005). It encompasses much more than the freedom to create one's schedule, and outcomes of an autonomous environment are highly interdependent among team members. In the context of spaceflight, autonomy refers to the extent to which the crew will act independently from mission control to complete objectives and/or respond to complications/emergency situations when needed due to environmental conditions (i.e., distance), as well as the extent to which the crew will prioritize mission objectives and schedule activities (Reagan and Todd, 2008).

Concerning informational risks, data on astronauts' health, both during and following missions, are potentially valuable resources for multiple purposes, including continuous learning about health risks (e.g., refining future health and safety standards for the benefit of future crews), internal assessment and quality improvement activities, and promoting transparency and public trust through evidence-based policies and decisions. Because there is only a small sample of participating astronauts, however, commonly used strategies for protecting individual privacy (e.g., de-identifying data) may not be effective. Even if overt identifiers are stripped, the data may be intrinsically identifiable if, for example, only two men were on a mission and two cases of post-mission prostate cancer were reported. Long-duration and exploration spaceflights present opportunities where access to health data can support activities with very high social value, but where privacy may be particularly difficult to protect. It is in the interest of current and future astronaut crews to enable appropriate uses of astronaut health data to aid continuous learning and improvement of safety standards and risk-mitigation strategies. However, stringent policies to protect the privacy and confidentiality of sensitive individual health information are an essential precondition for such activities.

Astronauts have the ability to report inflight medical problems via a private medical conference. However, it may be possible to increase the information being collected and address potential problems of underreporting of health problems by astronauts, if an anonymous reporting mechanism was implemented and its use encouraged.

FAIRNESS

The principle of fairness (or more appropriately justice as fairness) requires that like cases should be treated alike, or alternatively stated, equals should be treated equally and unequals should be treated unequally. In various social contexts, the principle of fairness has application to the distribution of goods and risks (distributive justice), righting past wrongs (compensatory justice), or assuring fair processes (procedural justice). One application of this principle notes that individuals, and identifiable groups, should be treated the same if they are in the same or similar circumstances unless there is a reasonable and ethically acceptable rationale for treating them differently. Thus, treatment does not have to be exactly the same, but rather must be fair or equitable. For example, in biomedical research, it makes little sense to enroll women into a trial to assess the efficacy of a new drug to treat prostate cancer—equality of the opportunity to participate is not required. However, fairness (equity) requires that there be clinical trials to assess the efficacy of new drugs to treat cancers that affect only women.

Putting the Principle in Context

Considerations of distributive justice underlie decisions about how the risks and benefits of research involving human participants should be shared, especially when particular individuals or groups experience the risks of research without the prospect of realizing its potential benefits. For example, healthy individuals exposed in toxicological studies bear risks from adverse effects for benefits that primarily accrue to society at large. In contrast, the history of women's exclusion from clinical research studies has forestalled significant benefits by reducing the available knowledge base for treatment of health conditions that affect women.

In addition, men were unfairly being burdened with the risks of such research (IOM, 1994).

Compensatory justice supports, for example, the inclusion of greater numbers of women in clinical research studies and the increased study of women's health conditions, in response to this historical exclusion. Compensatory justice also supports the government's provision of health benefits to veterans. As a nation, we ask those joining the military to undertake substantial risk, including the risk of death, for the benefit of the nation's collective national security.

Last, procedural justice requires fairness in processes, for example, in decision making about dispute resolution and resource allocation. The U.S. judicial system was designed to provide fair decision-making processes for dispute resolution, with uniform procedures requiring notice and opportunity to be heard to ensure that each side can present facts in support of a position, and with a right to appeal a decision a party deems to be unfair. Another example of procedural justice involved public participation in resource allocation decisions. When the state of Oregon was faced with difficult decisions on provision of health services in light of limited resources, it established a series of engagement opportunities for citizens to provide input into that decision (Dixon and Welch, 1991; Fleck, 1994).

Focusing on the Principle in the Context of Spaceflight

As noted throughout this chapter, one important ethical challenge of exposing humans to the risks of long duration and exploration spaceflight is that the burden of the health risks associated with these missions falls to a limited number of astronauts and their families while the benefits of the proposed missions accrue primarily to future astronauts and more broadly to society. In addition to being a concern of appropriate risk-benefit balance, the appropriate risk-benefit distribution must also be considered. Asking individuals to accept great risk (either in likelihood or magnitude of harm) can be partially balanced by making a commitment to provide long-term health care and health monitoring as is done for military veterans (see further discussion in the next section on fidelity). Distribution of risks within the crew will need to be as fair as possible given that some jobs (e.g., those involving extravehicular activities) might put some team members at higher risk than others.

Concerns about fairness also focus on individual and group susceptibilities to risk, as well as to fairness issues in crew composition. Issues have arisen specifically about risks of radiation-induced cancers being higher for women (see Chapter 3). Whereas historically concepts such as "concerns over reproductive issues for women" have been used to exclude women from certain occupations,[5] the solution was, ultimately, not to exclude women but to make the workplace safer. A rational system would acknowledge individual susceptibilities across a wide spectrum of exposures and use the best available knowledge to ameliorate the effects. In the case of women's participation in long duration and exploration space missions, excluding women from the early missions (as might happen, for example, due to sex differences in lifetime limits on permissible exposure levels to radiation [see Chapter 3]) creates an unfairness that will persist and become self-fulfilling over time. Less information will be available about the health effects of participation by women, leading to greater uncertainty about the risks to women as compared to men over the course of multiple missions and, in the process, undermining equitable opportunities for women to participate. Unless women participate in missions, important information cannot be collected, which affects the implementation of other important principles, most notably those that require avoiding harm and appropriately balancing risk and benefit.

Similar principles apply to NASA's ongoing efforts to ensure diversity in crew selection and create true equality of opportunity to participate in long duration and exploration spaceflight. As more information emerges about biological markers for susceptibility to disease, it is possible that virtually every astronaut candidate will be found to have one or more marker of increased sensitivity. For example, one individual might be more sensitive to radiation, another to bone loss during microgravity, another to the psychological effects of isolation and confinement in close quarters, and another to visual changes associated with microgravity. Fairness requires a holistic consideration of those diverse sensitivities with special attention to equitable distribution of risks and benefits and equality of opportunity.

---

[5]For example, in the 1980s, a battery manufacturing company put in place a policy that denied women of child-bearing capacity the opportunity to work on jobs involving potential exposure to lead. The U.S. Supreme Court ultimately ruled that such policies violated Title VII of the Civil Rights Act of 1964 (*Automobile Workers v. Johnson Controls, Inc.*, 499 U.S. 187 [1991]).

FIDELITY

In situations where risks are to some degree unquantifiable, uncertain, and unknowable, and so cannot be well managed in advance, the principle of fidelity has been proposed as a "promise to stand by after" (Zoloth, 2013). Those who consent to incur long-term health risks for society's benefit are entitled to fidelity, reflected in society's durable commitment to minimize any harms that emerge, whenever they emerge. This concept of fidelity or reciprocity resonates with the basic, widely shared understanding that it is unjust to allow "some people alone to bear public burdens which, in all fairness and justice, should be borne by the public as a whole."[6] As a practical matter, the public cannot physically share the risks that astronauts will bear. It can, however, share the costs and burdens of ongoing risk-mitigation efforts. The astronaut's consent to participate in the face of uncertain risks gives rise to a mutuality of obligation, akin to the legal concept of "future consideration" in which one party's performance gives rise to duties that are owed (Garner, 2009).

Putting the Principle in Context

In most cases, persons who consent to work in hazardous environments in terrestrial settings have the option to stop if risks become unacceptable. Ethical and regulatory frameworks for research involving human participants similarly recognize the right of those individuals who volunteered for the research study to withdraw consent to participate at any time (45 C.F.R. 46.116(a)(8)). When consent is revocable, there is a reduced ethical imperative to define clear substantive duties that are owed to those who agree to participate; withdrawal is nearly always an option if they grow dissatisfied with how they are treated or find the conditions of participation unacceptable. Of course, there is a duty to avoid harm and minimize the risk of research protocols. Beyond this general requirement, there has been discussion that the rights of human research participants imply duties on the part of research sponsors and investigators (Faden and Beauchamp, 1986). However, it has been difficult to forge consensus about concrete, substantive duties—such as a duty to provide care or reimburse costs for research-related injuries—that sponsors and investigators may owe research participants.

---

[6] *Armstrong v. United States*, 364 U.S. 40, 49 (1960).

The nonbinding nature of research participation has been seen as blunting the need to develop a concept of fidelity that imposes binding, affirmative obligations owed by research sponsors and investigators to a consenting research volunteer.

Focusing on the Principle in the Context of Spaceflight

An astronaut's consent becomes binding and irrevocable at the moment the mission launches. Astronauts are free to withdraw their agreement to participate prior to launch, but from the launch moment forward, it becomes near impossible to turn back, and astronauts likely will encounter uncertain and unquantifiable risk exposures and endure potential harms to health that will persist after the mission.

The irrevocability of participation, once begun, in long duration and exploration spaceflight creates an ethical imperative to define long-term duties owed to the participating astronaut. This is a necessary corollary of the ethical principle of avoiding or removing harm, and can be further supported by the principle to provide benefit. In this context the principles support the minimization of risk of harm, the treatment of injuries or health conditions during the flight, and the ongoing monitoring and provision of health care after the flight. This binding duty to provide ongoing surveillance, monitoring, and health care during the lifetime of the astronaut is part of the continuum of risk management that begins with engineering and design efforts to minimize risk and continues through the flight and postflight.

RECOMMENDATION

Recommendation 2: *Apply Ethics Principles to Health Standards Development and Implementation*
NASA should apply the following ethics principles in the development and implementation of its health standards for decisions regarding long duration and exploration spaceflights:

- ***Avoid harm***—the principle includes the duty to prevent harm, exercise caution, and remove or mitigate harms that occur. Thus, NASA should exhaust all feasible measures to minimize the risks to astronauts from long duration and exploration spaceflights, including addressing uncertainties through approaches to risk prevention and mitigation that incorporate safety margins and include mechanisms for continuous learning that allow for incremental approaches to risk acceptance.

- ***Beneficence***—the principle to provide benefit to others. NASA should consider in its decision making the potential benefits of a specific mission, including its scientific and technological importance, as well as its potential beneficiaries including current and future astronauts and members of society at large.
- ***Favorable balance of risk and benefit***—the principle to seek both a favorable and acceptable balance between the risk of harm and potential for benefit. In authorizing long duration and exploration activities and in approving particular missions, NASA should systematically assess risks and benefits and the uncertainties attached to each, drawing on the totality of available scientific evidence, and ensuring that benefits sufficiently outweigh risks.
- ***Respect for autonomy***—the principle to ensure that individuals have both the right to self-determination and processes in place to exercise that right. NASA should ensure that astronauts are able to exercise voluntariness to the extent possible in personal decision making regarding participation in proposed missions, that they have all available information regarding the risks and benefits of the proposed mission, and that they continue to be apprised of any updates to risk and benefit information throughout the mission.
- ***Fairness***—the principle requires that equals be treated equally, that burdens and benefits be distributed fairly, and that fair processes be created and followed. NASA's decision making surrounding missions should explicitly address fairness, including the distribution of the risks and benefits of the mission, crew selection, and protections for astronauts after missions.
- ***Fidelity***—the principle recognizes that individual sacrifices made for the benefit of society may give rise to societal duties in return. Given the risks that astronauts accept in participating in hazardous missions, NASA should respect the mutuality of obligations and ensure health care and protection

for astronauts not only during the mission but after return, including provision of lifetime health care for astronauts.

REFERENCES

ALI (American Law Institute). 1979. Restatement of the law second, torts. Vol. 4. Philadelphia, PA: ALI.

ALI. 1981. Restatement of the law second, contracts. Philadelphia, PA: ALI.

Aristotle. 2009. *The Nicomachean ethics*, edited by L. Brown, translated by W. D. Ross. New York: Oxford University Press.

Beauchamp, T. L. 2003. A defense of the common morality. *Kennedy Institute of Ethics Journal* 13(3):259-274.

Beauchamp, T. L., and J. F. Childress. 2013. *Principles of biomedical ethics*. 7th ed. New York: Oxford University Press.

Bergelson, V. 2007. The right to be hurt: Testing the boundaries of consent. *George Washington Law Review* 75:165-236.

Dixon, J., and H. G. Welch. 1991. Priority setting: Lessons from Oregon. *Lancet* 337(8746):891-894.

Dworkin, G. 1988. *The theory and practice of autonomy*. Cambridge, UK: Cambridge University Press.

Faden, R. R., and T. L. Beauchamp. 1986. *A history and theory of informed consent*. New York: Oxford University Press.

Fallon, R. H. J. 1994. Two senses of autonomy. *Stanford Law Review* 46(4):875-905.

Feinberg, J. 1986. *Harm to self: The moral limits of the criminal law*. New York: Oxford University Press.

Fischhoff, B., S. Lichtenstein, P. Slovic, S. L. Derby, and R. Keeney. 1981. *Acceptable risk*. Cambridge, UK: Cambridge University Press.

Fleck, L. M. 1994. Just caring: Oregon, health care rationing, and informed democratic deliberation. *Journal of Medicine and Philosophy* 19(4):367-388.

Garner, B. A. 2009. *Black's law dictionary*. 9th ed. Eagan, MN: West Publishing Company.

Gert, B. 1998. *Morality: Its nature and justification*. New York: Oxford University Press.

HEW (Department of Health, Education, and Welfare). 1979. *The Belmont report: Ethical principles and guidelines for the protection of human subjects of research*. http://www.hhs.gov/ohrp/humansubjects/guidance/belmont.html (accessed January 2, 2014).

IOM (Institute of Medicine). 1994. *Women and health research: Ethical and legal issues of including women in clinical studies, Volume 1*. Washington, DC: National Academy Press.

IOM. 2001. *Safe passage: Astronaut care for exploration missions.* Washington, DC: National Academy Press.

IOM. 2009. *Beyond the HIPAA privacy rule: Enhancing privacy, improving health through research.* Washington, DC: The National Academies Press.

Kahn, P. W. 1992. *Legitimacy and history: Self-government in American constitutional theory.* Binghamton, NY: Vail-Ballou Press.

Kant, I. 1998. *Groundwork of the metaphysics of morals,* edited by M. J. Gregor. Cambridge: Cambridge University Press.

Law Commission of England and Wales. 1995. *Consultation paper no. 139: Consent in the criminal law.* http://www.bailii.org/ew/other/EWLC/1995/c139.pdf (accessed February 26, 2014).

Leach, D. J., T. D. Wall, S. G. Rogelberg, and P. R. Jackson. 2005. Team autonomy, performance, and member job strain: Uncovering the teamwork KSA link. *Applied Psychology* 54(1):1-24.

Mastroianni, A., and J. Kahn. 2001. Swinging on the pendulum. Shifting views of justice in human subjects research. *Hastings Center Report* 31(3):21-28.

Mill, J. S. 1879. *Utilitarianism.* 7th ed. London: Longmans, Green, and Co.

Moreno, J. D. 2001. Goodbye to all that: The end of moderate protectionism in human subjects research. *Hastings Center Report* 31(3):9-17.

Moskop, J. C., C. A. Marco, G. L. Larkin, J. M. Geiderman, and A. R. Derse. 2005. From Hippocrates to HIPAA: Privacy and confidentiality in emergency medicine—Part 1: Conceptual, moral, and legal foundations. *Annals of Emergency Medicine* 45(1):53-59.

NASA (National Aeronautics and Space Administration). 2010. *Research and technology development to support crew health and performance in space exploration missions.* NASA Research Announcment NNJ10ZSA003N. http://www.oorhs.pitt.edu/Documents/EO_2010_08_02_002.pdf (accessed January 31, 2014).

NASA. 2013a. *NASA's vision.* http://www.nasa.gov/about/index.html (accessed January 3, 2014).

NASA. 2013b. Astronaut candidate class. http://www.nasa.gov/astronauts/2013 astroclass.html (accessed February 6, 2014).

National Bioethics Advisory Commission. 1999. *Research involving human biological materials: Ethical issues and policy guidance.* Vol. 1. Pp. 71-72. https://bioethicsarchive.georgetown.edu/nbac/hbm.pdf (accessed February 6, 2014).

Rawls, J. 1993. *Political liberalism.* New York: Columbia University Press.

Raz, J. 1986. *The morality of freedom.* New York: Oxford University Press.

Reagan, M. and B. Todd, 2008. *"Autonomy" strategies and lessons from the NEEMO project.* Houston, TX: NASA Johnson Space Center.

Sandman, P. 1989. Hazard versus outrage in the public perception of risk. In: *Effective risk communication: The role and responsibility of government and nongovernment organizations*, edited by V. T. Covello, D. B. McCallum, and M. T. Pavlova. New York: Plenum Press. Pp. 45-49.

Shapiro, D. L. 1988. Courts, legislatures, and paternalism. *Virginia Law Review* 74(3):519-575.

Slovic, P. 1987. Perception of risk. *Science* 236(4799):280-285.

Smith, R. M. 1982. The constitution and autonomy. *Texas Law Review* 60(2):175-205.

Tauber, A. I. 2005. *Patient autonomy and the ethics of responsibility*. Cambridge, MA: MIT Press.

Zoloth, L. 2013. *Uncertainty, testimony, and fidelity: Ethical issues in human space exploration*. PowerPoint presented at the second meeting of the IOM Workshop Committee on Ethics Principles and Guidelines for Health Standards for Long Duration and Exploration Spaceflights, Washington, DC, July 25. http://www.iom.edu/~/http://www.iom.edu/~/media/Files/Activity http://www.iom.edu/~/media/Files/Activity%20Files/Research/HealthStandards Spaceflight/2013-JUL-25/Panel%203%20-%20Zolof%20FINAL%20FINAL% 20Ethical%20Issues%20in%20Human%20Space%20Explo.pdf (accessed February 6, 2014).

6

Recommendations for Ethics Responsibilities and Decision Framework

An ethics framework to guide decision making is more than just the identification of applicable, individual ethics principles. The translation of ethics principles into ethical policy generally requires specific actions by specific actors across multiple levels, including individual, administrative, and societal. As the National Aeronautics and Space Administration (NASA) considers how to approach ethical questions associated with health standards for long duration and exploration spaceflight, NASA and other policy makers must confront the additional responsibilities that accompany decisions allowing individuals to accept levels of risk to health and safety beyond those generally acknowledged as upper limits. This chapter discusses specific responsibilities that stem from the ethics principles identified in Chapter 5, and endorses a decision framework for health standards for long duration and exploration spaceflight. The recommendations in this chapter contain many components that NASA has considered or is already implementing in its work on health standards for spaceflight. However, the adoption of the recommended responsibilities and decision framework not only introduces new processes and responsibilities but also alters the context in which existing components operate.

ETHICS RESPONSIBILITIES FOR LONG DURATION AND EXPLORATION SPACEFLIGHTS

As an employer, a federal agency responsible for innovation and exploration, a research sponsor, and an international partner, NASA has moral obligations to formally recognize and act on responsibilities that logically flow from the ethics principles outlined in Recommendation 2 if long duration and exploration spaceflights are acceptable. Specifically, the ethics principles need to be incorporated into decisions about whether risks in excess of those allowed under current health standards are acceptable, and, if so, what conditions must be satisfied to engage in ethically acceptable long duration and exploration class missions.[1] The committee's view is that such responsibilities are consistent not only with the principles identified but also with a range of ethical theories, without any intended or unintended endorsement of a particular ethical theory.

Recommendation 3: *Implement Ethics Responsibilities*
NASA should adopt policies or processes that formally recognize the following ethics responsibilities related to health standards for long duration and exploration spaceflights:

- **Fully inform astronauts about the risks of long duration and exploration spaceflights and make certain that the *informed decision-making process* is adequate and appropriate.**
- **Adhere to a *continuous learning strategy* (including health surveillance and data collection) to ensure that health standards evolve and improve over time and are informed by data gained before, during, and after long duration and exploration spaceflights, as well as from other relevant sources.**
- **Solicit *independent advice* about any decision to allow any specific mission that fails to meet NASA health standards or any decision to modify health standards.**
- **Communicate with all relevant stakeholders (such as astronauts and the public at large) the rationale for, and possible impacts (including harm type, severity, and probability estimates) related to any decision about health standards in a *procedurally transparent, fair and timely manner*, providing adequate opportunity for public engagement.**

---

[1]Limitations imposed by costs, time lines, and technological feasibility also influence decisions about long duration and exploration spaceflights.

- **Provide *equality of opportunity* for participation in long duration and exploration spaceflights to the fullest extent possible. For example, fairness in crew selection means that NASA should accept some group differences in population risk in order to create equality of opportunity to participate in missions, and accommodate individual variance from population-based risk estimates to the extent that individual differences do not jeopardize mission operations.**
- **Provide *preventive long-term health screening and surveillance* of astronauts and *lifetime health care* to protect their health, support ongoing evaluation of health standards, improve mission safety, and reduce risks for current and future astronauts.**
- **Develop and apply policies that appropriately and sufficiently *protect the privacy and confidentiality* of astronaut health data.**

The committee recognizes that ethics principles would rarely be considered or applied in isolation when making decisions about health standards for long duration and exploration spaceflights. The following section discusses a few examples of how ethics responsibilities may overlap or can be applied in a coordinated approach.

Informed Decision Making

The need for well-informed decisions involving risk is recognized in many different legal and ethical contexts, including separate but related informed consent requirements for research involving human participants and for medical treatment decisions (Faden and Beauchamp, 1986; CIOMS, 2002; Mazur, 2003; Tauber, 2005); securities regulations that rely on mandated disclosures by companies as the primary mode of risk protection for investors (Schwartz, 2010); and diverse legal doctrines that treat consents and agreements as ineffective if parties did not understand the acts and consequences to which they were agreeing (e.g., ALI, 1979). Considerable consensus exists about certain requirements that must be satisfied for an individual's consent to be valid:

- Decisional competence of the consenting individual (e.g., an adult of sound mind)

- Voluntary decision making (i.e., free of undue influence or pressure)
- Sufficient understanding and comprehension of the elements involved (e.g., nature, duration, purpose, and risks) to make an informed decision about the consequences of consent. (Faden and Beauchamp, 1986; Gostin, 1995; Emanuel et al., 2000)

The committee adopted the term "informed decision making" to refer to the process through which astronauts make individual decisions to participate in specific long duration and exploration spaceflight. In many respects, astronauts' participation during such spaceflight parallels other contexts in which the concept of informed consent is applied, especially when space missions involve human research. However, important distinctions exist that make the term "informed consent" inappropriate in the context of this committee's task.

Sophisticated Decision Makers

The concept of "informed consent" as applied in medical settings and in research involving human participants presumes significant information asymmetries between knowledgeable clinicians or investigators and the non-expert (and often ill) patients or research participants who must consent to treatment or participation. However, courts have drawn distinctions between different types of decision makers in the context of informed decision making. For example, courts distinguish between unsophisticated and sophisticated investors when applying the protections of U.S. securities regulations (Fletcher, 1988). Even in this circumstance, courts afford some protection to the sophisticated investor, particularly with respect to ensuring accurate disclosure of risk information: "Sophisticated investors, like all others, are entitled to the truth."[2]

Similarly, concerns about lack of sophistication and inability to understand and apply information are usually more limited in the context of astronauts' decisions about spaceflight risks. Astronauts are members of a professional community and culture committed to discovery and exploration, and their role is, in part, that of a co-venturer who shares (rather than passively incurs) risk in pursuit of goals that are personal as well as organizational. As part of the astronaut corps training, NASA provides

---

[2] *Stier v. Smith*, 473 F.2d. 1205, 1207 (5th Cir. 1973).

astronauts with detailed risk information during all phases of their training and work, including regular updates and involvement of the astronauts in all phases of the analysis and dissemination of risk data (Behnken, 2013). These efforts proactively afford astronauts an opportunity to consider the risks somewhat dispassionately, and over extended time periods, before accepting a crew position for any given spaceflight.

Decisions to participate in long duration and exploration spaceflights must be consistent with valid informed decision making for individuals engaged in activities with elevated levels of risk. Not all risks can be morally adjudicated by leaving the decisions entirely to high-functioning, autonomous, and sufficiently voluntary participants. Astronauts are occupationally, psychologically, and culturally predisposed to accept the risks associated with spaceflight. NASA already recognizes the need for additional layers of approval in situations involving elevated levels of risk. In addition to an astronaut's decision to participate in a specific mission, the "responsible program, project, or operations manager needs to formally accept" residual risks to human safety, in addition to formal concurrence from the responsible technical authorities (NASA, 2012, p. 35).

Concerns about possible influences on individual astronaut's decisions to agree or refuse participation in activities with certain types or levels of risk persist. After years of training, astronauts may feel pressured to accept residual risk[3] or risks inherent to human research during the mission because selection for spaceflight is rare. NASA procedural requirements state that an astronaut's decision not to participate in research during a spaceflight "may result in removal of that individual from that mission" (NASA, 2008, p. 20). Although the procedural requirements state that such withdrawal should not influence an astronaut's future career opportunities, the withdrawal may be considered in decisions about the astronaut's inclusion in other missions with similar research protocols (NASA, 2008).

Irreducible Uncertainty of Available Risk Information

Another decisional challenge lies in "confronting the unknown and making difficult choices in the face of limited information" (Farber, 2011, p. 901). The reality of long duration and exploration human spaceflight is that there is an irreducible level of uncertainty about the risks

---

[3]Residual risk "is the remaining risk that exists after all mitigation actions have been implemented or exhausted in accordance with the risk management process" (NASA, 2013, p. 2).

involved. Even after disclosure of the best available information about risks to the decision maker, a truthful and complete disclosure must acknowledge that material risks may exist that remain unforeseen and unknown, and many risks that are anticipated are impossible to quantify with any precision. Informed decision making does not require that all risk factors be known or quantifiable, but efforts to communicate the uncertainties to astronauts are a fundamental part of acting on ethics principles in the context of long duration and exploration spaceflight.

Continuous Learning

As noted throughout this report, long duration and exploration spaceflights include some highly uncertain and unknown risks, particularly those associated with the types and probabilities of harm. Implementing the relevant ethics principles requires the collection and analysis of information before, during, and after the mission to reduce uncertainty and improve understanding of known health and safety risks.

Continuous learning in health care has the potential to improve the care of individual patients, as well as other future patients, by rapidly translating data collected from patients' encounters into generalizable knowledge that can be incorporated into future clinical practice (IOM, 2007). The "learning health system" model is based upon the implementation of effective and efficient policies to promote continuous learning. The Institute of Medicine (IOM) has defined a learning health system as a health care system "in which knowledge generation is so embedded into the core of the practice of medicine that it is a natural outgrowth and product of the healthcare delivery process and leads to continual improvement in care" (IOM, 2007, p. 6).

This model has had profound implications for clinical and research ethics involving humans because ethics and policy in this area have traditionally recognized a sharp distinction between the clinical care of patients and research that involves human participants (Kass et al., 2013). Efforts are now under way to modernize ethical frameworks in response to this blurring of roles (Faden et al., 2013).

Continuous learning has clear application for long duration and exploration missions, including the collection of information for both monitoring and research on individual health impacts and for setting or refining health standards. As NASA's own processes for continuous risk management indicate, each space mission provides opportunities for learning about health risks, thus enabling refinement of monitoring, estimations, and management. In addition to these efforts, NASA has opportunities to strengthen its policies about collection and analysis of data relevant to health risks and uncertainties. In a recent report, the IOM noted a huge gap in knowledge of how medications are used during spaceflight because "there is no mandatory systematic collection of data" (IOM, 2014, p. 15). The IOM has suggested the use of an occupational health and safety model to enhance data collection at NASA (IOM, 2001, 2014). As part of a policy for managing health risks and uncertainty, NASA should build on its health monitoring processes and protocols before, during, and after missions (including its longitudinal study of astronauts' health) with the goal of learning more about the short- and long-term health effects of spaceflight on astronauts and the impact of prevention and treatment strategies. Often, uncertainties can be diminished with relatively modest exposures to risk, and policies enabling continuous learning could contribute to substantial advancements in understanding of, and responses to, risks and uncertainties related to current and future spaceflights involving humans.

Of course, data collection often implicates important privacy and confidentiality interests that must be appropriately protected and respected. A 2007 IOM report on the learning health care system advocated "advancement … of the use of clinical data as a central common resource for advancing knowledge and evidence for effective care—including directly addressing current challenges related to the treatment of data as a proprietary good and interpretations of the Health Insurance Portability and Accountability Act (HIPAA) and other patient privacy issues that currently present barriers" (IOM, 2007, p. 6). Concerns about confidentiality may affect an astronaut's willingness to supply medical data for purposes beyond current treatment. However, such cooperation will likely serve astronauts' interests when shared medical information is used to refine treatment or protocols during the course of a mission or to revise health standards for future missions. The use of a "trusted broker" as an intermediary to ensure patient confidentiality while facilitating access to data is one possible means of ensuring privacy while collecting useful data (Boyd et al., 2009). More nuanced balancing of the goals of privacy protections and continuous learning will be required when the information about individuals sought is: (1) sensitive (i.e., with potential negative impact for future participation in missions, misuse by others, or even embarrassment); and (2) likely to be used in ways that will benefit astronauts on future missions rather than the individuals from whom the information is sought.

Independent Advice

In the United States, independent advisors or decision makers are commonly involved in regulatory activities or policy determinations that involve high-risk activities. Many institutions charged with engaging in risky activities are also responsible for related risk management assessments. For instance, U.S. academic institutions, which earn revenue and prestige by hosting clinical trials, also are required to perform their own ethical assessments of human studies. NASA is similarly structured. According to the National Aeronautics and Space Act, NASA's responsibilities include, among others, the expansion of human knowledge of space through experience and research, as well as improvement of the efficiency and safety of space vehicles.[4] To put it simply, NASA is obligated to engage in space travel while managing health and safety risks, technological limitations, costs, and time lines associated with overarching programs and specific missions. Such a governance structure has the potential to lead to significant conflicts of interest within the governing agency or institution.

In many other high-risk fields, independent review processes are used to insulate decisions about safety and risk from other competing institutional imperatives, such as production pressures. For example, in research involving humans, institutional review boards review, approve, and may modify research protocols independently of, but in coordination with, research-related entities to ensure compliance with federal research rules and regulations and the ethics principles intended to protect research participants (HHS, 1993). In some cases, elevated risks to research participants may require oversight by Data Safety Monitoring Boards, which review and evaluate study data to maintain participant safety, study conduct and progress, and efficacy (NIDCR, 2014).

Independent advisors should inform NASA's decisions about whether to participate in long duration and exploration spaceflights that are unlikely to meet existing health standards to protect astronauts' health and safety. NASA's process to establish health standards involves both internal and external medical experts (see Chapter 2). Before determining whether specific circumstances allow for spaceflights that fail to meet health standards, NASA should continue to seek independent advice from individuals with relevant medical, social, ethical, and engineering expertise (including former astronauts). Further, such independent analy-

[4]National Aeronautics and Space Act of 1958, P.L. 85-568 (July 29, 1958).

sis should include people with varying but credible views on mission benefits, risks, and safety countermeasure feasibility to ensure a thorough and exhaustive analysis before NASA reaches any final decisions.

Transparency

Successful long duration and exploration missions require sufficient levels of transparency regarding policy and procedure development and implementation to promote accountability, provide a mechanism and incentive for amending decisions, and build trust among stakeholders. Although transparency can also carry liabilities, such as inhibition of open and frank deliberations, the committee views transparency as a responsibility to share information and engage external stakeholders, including the public, as critical voices to shape the vision and sustain the moral foundation of public institutions.

The committee endorses NASA's recent efforts to enhance its decision-making culture through processes that are available to all stakeholders, which is an important step toward building and maintaining the ongoing trust in NASA as an agency and in its directives. As a public agency whose actions are of great interest and potential consequence to the broad public, NASA will be more likely to satisfy its ethics responsibilities if opportunities are provided for public comment and input about decisions related to health standards for long duration and exploration spaceflight.

The *Federal Register* is an effective means by which to engage the public, and NASA should consider using this vehicle not only to maintain transparency of process but also to seek useful suggestions, advice, and reactions from the public about health standards for long duration and exploration spaceflights. Many agencies use the *Federal Register* to post notices of proposed and final rulemaking and to provide the public with an opportunity to review and comment on government regulations or policies. For example, proposed new or modified OSHA regulations are subject to public review and comment as a statutory requirement.

Under the National Aeronautics and Space Act, in performance of its functions to "plan, direct, and conduct aeronautical and space activities," NASA is authorized to make rules and regulations governing the manner of its operations.[5] Pursuant to the Administrative Procedure Act, which exempts matters relating to agency management or personnel from *Fed-*

[5]51 U.S.C. 203(a)(1).

eral Register publication requirements,[6] NASA is not required to publish its health standards in the *Federal Register* for formal rulemaking.[7] In most cases, it would be possible for NASA to use the *Federal Register* in a way that would not needlessly delay or interfere with sound and timely decisions. The committee believes that using mechanisms such as the *Federal Register* to enhance transparency about decisions regarding health standards for long duration and exploration spaceflights would increase understanding about motivation, goals, and objectivity and promote stakeholder trust and willingness to support the commitment of resources required for long duration and exploration missions.

AN ETHICS-BASED DECISION FRAMEWORK

If a human spaceflight mission cannot meet current health standards, or if inadequate information exists to revise a health standard, the options, as identified and examined by the committee, would be to: (1) liberalize the NASA health standards, (2) establish more permissive "long duration and exploration health standards," or (3) grant an exception to the standard in order to conduct these missions before new protective technologies and strategies are available or additional data are acquired which may allow revision of the standard. The committee found the first two options to be unacceptable when evaluated against the ethics principles and responsibilities described in this report.

The committee finds relaxing (or liberalizing) current health standards to allow for specific long duration and exploration missions to be ethically unacceptable. Current NASA policy outlines the administrative processes and levels of approval required to initiate a new health standard or revise a current health standard. NASA's health standards "are based on the best available scientific and clinical evidence, as well as operational experience" (NASA, 2007, p. 8). Moreover, review of health standards occurs every 5 years or can be triggered any time that new research data or clinical observations indicate an update may be necessary. Modifying health standards outside of this established process merely to permit long duration and exploration missions would be arbitrary.

The second option maintains current health standards for all missions except long duration and exploration spaceflight, which would have a

---

[6] 5 U.S.C. 553(a)(2).

[7] Personal communication, Richard Williams, NASA, March 6, 2014.

separate set of health standards. This approach presumably would entail setting a more permissive ceiling on allowable risk for long duration and exploration missions under conditions in which existing evidence and knowledge make it nearly impossible to quantify those ceilings. The committee found this approach wanting, lacking a clear and compelling justification for why acceptable risks and levels of uncertainty should be greater for long duration and exploration missions than other human spaceflight missions.

Having excluded the options of modifying existing standards or creating a separate set of standards, the committee concludes that the only ethically acceptable option that could allow for increased risk exposures in the context of long duration and exploration spaceflights is granting an exception to existing health standards. The committee believes that exceptions to health standards should be considered on a mission-by-mission basis and used under very limited circumstances following the ethics-based decision framework recommended. To determine whether exceptions to health standards are ever acceptable for space missions and, if so, whether individual missions adhere to the ethics principles and responsibilities articulated in this report, NASA should adopt the following ethics-based decision framework based on three separate levels of decision making. In addition to concerns about astronaut health and safety, considerations about cost, technical feasibility, and timeliness will also influence these decisions, as they affect both potential risks and benefits of specific missions.

Recommendation 4: ***Adopt an Ethics-Based Decision Framework***
NASA should apply the relevant ethics principles and fulfill the concomitant responsibilities through a three-level, ethics-based decision framework that examines

- **Level 1: Decisions about allowing risk to astronaut health and safety in excess of that permitted by health standards,**
- **Level 2: Decisions about undertaking specific missions, and**
- **Level 3: Decisions concerning individual astronaut participation and crew composition.**

The first and broadest decision (Level 1) is whether, and under what conditions, any missions that are unlikely to meet current health standards are ethically acceptable. If such missions are ethically acceptable, NASA must articulate specific conditions that must be fulfilled to approve an exception to health standards. If NASA decides that missions that fail to meet existing health standards are not acceptable, then such missions must be deferred until new knowledge about risk or uncertainties or risk mitigation strategies are available.

The second level of decision making concerns whether a specific mission is ethically acceptable. This analysis requires evaluating whether the mission meets the conditions enumerated under Level 1. Whereas Level 1 could be a one-time decision by NASA that affects all long duration and exploration missions that are unlikely to meet existing health standards,[8] Level 2 decisions would be made on a mission-by-mission basis.

If a specific mission is deemed ethically acceptable, the third level of decision making involves decisions by both NASA and individual astronauts. Level 3 decisions include, among other considerations, identification of mission crews given the requisite skills and expertise, the astronauts' health susceptibilities and personal risk factors (if known), and an individual astronaut's informed decision to participate. The following section describes how the recommended ethics-based decision framework could be applied.

Level 1: Decisions About the Limits of Health Standards

Level 1 decisions involve two basic questions (see Box 6-1). First, as a general rule, should NASA conduct space missions that will (a) fail to meet health standards, (b) involve significant risks where there are no applicable standards, and/or (c) involve such great uncertainty that NASA cannot exclude the possibility of a or b? Second, if so, what criteria should be used to determine whether exceptions for specific missions should be allowed?

Ethically, a thorough risk-benefit analysis is essential to determine whether health standards should ever be granted. The risks and uncertainties inherent to long duration and exploration spaceflights need to be balanced against the societal value or necessity of such missions, duties

[8]It is important to note that this decision may be revisited by NASA but should adhere to the same ethics-based decision framework to justify any alterations.

BOX 6-1
Decision Framework–Level 1:
Decisions About Missions That Fail to Meet Health Standards

Decision points:

- Should NASA conduct space missions that will (a) fail to meet health standards, (b) involve significant risks where there are no applicable standards, and/or (c) involve such great uncertainty that NASA cannot exclude the possibility of a or b?
- If so, what criteria should be used to determine whether exceptions for specific missions should be allowed?

Ethics principles and applications: Avoid harm, beneficence, acceptable risk-benefit balance, fidelity, transparency of decision making, commitment to continuous learning, procedural fairness of decision making

Examples of ethics responsibilities:

- Ensure that all feasible means will be taken to reduce astronaut risks to the lowest achievable levels
- Examine all approaches to minimizing risk including alternate approaches to meeting the mission's objectives
- Assess and communicate the benefits
- Determine and communicate the time urgency to conduct the mission
- Thoroughly monitor and conduct research on health impacts during and after spaceflight to inform current and future missions
- Commit to the future health of current and future astronauts by ensuring access to health care, longitudinal follow-up, and preventive screenings

to astronauts as expressed through fidelity, and a commitment to continuous learning and procedural fairness. When considering whether risks in excess of those allowed under existing health standards are ethically acceptable for long duration and exploration spaceflight, NASA should consider factors such as whether alternate approaches to obtaining desired information (e.g., unmanned spaceflights) are possible, whether delays in human spaceflights would likely lead to missions of similar benefits at significantly lower risk in the future, whether all feasible measures have been taken to minimize risks through engineering or other controls, and whether such missions are sufficiently urgent to justify risks to astronauts' health and safety above existing limits.

Two of the ethics responsibilities that result from sanctioning high-risk activities include continuous learning and engagement in health-related activities that protect astronaut health, support ongoing evaluation of health standards, improve mission safety, and reduce risks for current and future astronauts. Employers that knowingly expose employees to risks have an ethical responsibility to provide protection to the extent possible and to address the harms that occur when protection fails or turns out to be inadequate. Robust research and health-monitoring or surveillance programs that fully inform all who are involved, including astronauts and their families, are required. Furthermore, the committee maintains that the nation, through NASA, has the ethical duties to protect and sustain astronauts' health based on the ethics principles of fairness and fidelity. In the same way that the nation maintains its fidelity to its military veterans, providing lifetime health care to astronauts respects the commitment that astronauts have made and the risks they have taken on society's behalf.

If NASA decides that exceptions to existing health standards are ethically acceptable, in general, the second part of a Level 1 analysis requires NASA to decide what process and criteria should be applied to determine whether a specific mission should be granted an exception. An ethically acceptable process to grant exceptions would require a strict set of evidence-based criteria to evaluate whether a mission should qualify for an exception. The same criteria would apply to any mission that would fail to meet existing health standards. Based on the ethics principles identified, criteria for reviewing exception requests could include requirements that the proposed mission:

- be expected to have exceptionally great social value;
- have great time urgency;
- have expected benefits that would be widely shared;
- be justified over alternate approaches to meeting the mission's objectives;
- establish that existing health and safety standards cannot be met;
- be committed to minimizing harm and continuous learning;
- have a rigorous consent process to ensure that astronauts are fully informed about risks and unknowns, meet standards of informed decision making, and are making a voluntary decision; and
- provide health care and health monitoring for astronauts before, during, and after flight and for the astronaut's lifetime.

It is important to finalize the process and criteria prior to consideration of any specific mission to better ensure objective evaluation and selection of proposed criteria and to allow sufficient time for NASA to engage in an independent review of the selected process and criteria, thereby ensuring both objective and broad input about engagement in high-risk activities and associated risk management decisions. The committee emphasizes that, even if determined to be ethically acceptable, exceptions to existing health standards should only be granted in rare circumstances and should increase the responsibilities of NASA and society.

The selection of the process and criteria to grant exceptions to existing health standards should be evidence-based and should reflect policies that encourage independent advice and transparency of process to assuage and respond to stakeholder concerns about decisions that may expose U.S. astronauts to greater health and safety risks. The *Federal Register* would provide a central location for NASA to either solicit public input or publish the rationale and final decisions about whether to grant exceptions to existing health standards and, if so, about the process and criteria used to weigh exceptions for a specific long duration and exploration mission. Whether provided through the *Federal Register* or NASA's website, these notices or solicitations for public comment should define key terms (e.g., what constitutes an "exceptional" mission and what constitutes a "sufficient" understanding of relevant facts for purposes of informed decision making), identify common characteristics of the class of spaceflights at issue, designate specific ethics principles upon which analyses are based, explain how potential conflicts between principles were resolved, clarify the factors considered when weighing anticipated benefits and risks of harm, and describe any limitations to the decision's applicability.

Level 2: Decisions Regarding Specific Missions

If NASA concludes that any space missions that fail to meet health standards are ethically acceptable, Level 2 decisions focus on the health risks of specific, contemplated missions (see Box 6-2). If a mission fulfills the criteria established under Level 1, the scope of the health-standards exception (e.g., upper limits of excess exposure) would be determined on a mission-by-mission basis.

Decisions regarding a specific mission's goals and destination (e.g., the International Space Station for 12 months, the Moon, near-Earth asteroid, Mars) involve a complex evaluation of whether a mission meets the criteria articulated in the first tier of decision making. In particular, an independent ethical review of the specific mission should examine all processes to ensure that the relevant ethics responsibilities are implemented before, during, and after flight and that the public has access to decisions and actions related to those responsibilities. Given the long lead time of most mission planning, the ethical review process should include periodic reassessments, as warranted, during the period from provisional approval to launch. Poor decisions often result from excessive commitment to a prior decision without the benefit of subsequently available information, contrary to the ethics responsibility of continuous learning.

The ongoing ethics review would consider the risk-benefit balance, detail the feasibility of alternative mission designs, demonstrate that existing health standards still cannot be met despite technological advances and having taken all possible measures to reduce exposures to the lowest achievable level, and explore how this specific mission could impact the institutional integrity of NASA as a leader in safe and productive space exploration. The review would examine the plans for continuous learning throughout the flight and postflight, including how knowledge gained would be used to improve crew members' health, as well as to inform health standards for future crews. Additionally, the review would ensure that protocols for health status monitoring, treatment, and health care were in place postflight for astronauts.

BOX 6-2
Decision Framework–Level 2:
Decisions About Specific Missions

Decision point: Given authorization for missions that will likely fail to meet existing health standards, is a specific long duration and/or exploration mission ethically acceptable?

Ethics principles and applications: Avoid harm, acceptable risk-benefit balance, transparency of decision making, commitment to continuous learning, procedural fairness of decision making

Examples of ethics responsibilities:

- Adherence to criteria that are established and transparent
- Share risk escalation decisions and strategies
- Continue independent input to standards development and refinement
- Implement a robust program of occupational health monitoring and data collection during and after the mission
- Demonstrate that standards cannot be met despite having taken all feasible measures to reduce exposures to the lowest achievable level

Because NASA's health standards reflect "the best available scientific and clinical evidence" (NASA, 2007, p. 8), decisions about whether a specific mission is ethically acceptable should use existing health standards to characterize, to the extent possible, the levels of risk to astronaut health and safety beyond generally accepted levels of risk associated with specific long duration and exploration spaceflights. As such, NASA should continue to review emerging evidence to ensure appropriate revision of existing health standards and development of new health standards. NASA should also continue to identify gaps in knowledge and launch research efforts to generate information where needed.

Level 3: Decisions on Crew Selection and Individual Astronaut Participation

Level 3 includes decisions by NASA officials and individual astronauts related to crew selection and participation (see Box 6-3). NASA should continue its efforts to fully inform potential crew members about the nature and extent of spaceflight risks, the state of research knowledge, and the risk management plans. Ensuring that astronauts have the full range of information available on which to base their participation decisions (including information about unknown and uncertain risk) continues to be an essential component of NASA's risk management process. If an individual astronaut is invited to participate in space missions that expose that individual to risks in excess of those allowed by existing health standards, NASA must fully disclose the types, magnitudes, and possible consequences of known and uncertain risk beyond established limits, as well as the increased likelihood of unknown risks to health and safety.

As discussed in Chapters 4 and 5, informed decision-making processes relevant to long duration or exploration spaceflight participation may differ from traditional informed consent (e.g., clinical trials for drug development).[9] For example, information asymmetries between astronauts and NASA decision makers may not be as pronounced as between patient and doctor or research participants and clinical researcher. Moreover, astronauts' incentives to participate typically do not include the possibility of direct, physical benefit, and the consequences of not participating are either actually or perceived to be career limiting. Despite

---

[9]Note that traditional informed consent, in accordance with federal regulations, will continue to be needed for participation in research protocols.

these differences, efforts to insulate astronauts from actual or perceived undue pressure to participate in long duration and exploration spaceflights are essential.

In addition to individual astronauts' decisions about whether to participate in long duration and exploration spaceflights, Level 3 also requires NASA to make decisions about the balance between individual risk susceptibilities affecting health and the need for a diverse mission crew. The rare opportunities for spaceflight heighten the need to approach each mission as a source of potential data that may contribute to understanding of general health impacts of long duration and exploration spaceflight, and the risks for specific subpopulations. In the same manner that terrestrial medicine is rapidly recognizing individual variation in response to illness and medication, specific susceptibilities of individual astronauts should influence preventive measures, countermeasures, and tailoring of treatments. As more information emerges about genomic and other markers of susceptibility to injury or disease, it could be possible

BOX 6-3
Decision Framework–Level 3:
Decisions About Crew Selection and Individual Astronaut Participation

Decision point: What factors should be considered as NASA and individual astronauts make informed decisions about crew selection and individual astronaut participation for a given mission?

Ethics principles and applications: Informed decision making by astronauts, fairness, avoid harm, risk minimization (including risk to other crew members), commitment to continuous learning while protecting privacy and confidentiality

Examples of ethics responsibilities:

- Thorough sharing of risk data with astronauts
- Transparent and fair processes and policies on decision making
- Astronaut responsibilities to participate in data collection and health monitoring during and after spaceflights to inform current and future crews
- Selecting crew members in a manner that ensures fairness among groups and considers risk susceptibilities in general and for individuals in a way that allows inclusion, individual decision making within a range of risk that is prudent for the welfare of all astronauts during the mission

that every astronaut candidate would be found to have one or more marker of increased sensitivity. Depending on the long-term vision for human space travel, it will likely be important to understand the impact of spaceflight and the space environment on individuals with varied sensitivities and susceptibilities. Regardless of the balance struck between the ethics principles "avoid harm" and "fairness," NASA must follow polices and guidelines that ensure continuous learning while protecting astronauts' privacy and confidentiality.

CONCLUDING REMARKS

From its inception, human spaceflight has pushed the boundaries of acceptable health and safety risks for astronauts. NASA, its international partners, and commercial space ventures will continue to face complex ethical decisions about risk acceptability as technologies improve and longer and more distant spaceflights become feasible. The ethics framework outlined in this report builds on the work of NASA and others and identifies a set of recommendations for ethically assessing and responding to the challenges associated with health standards for long duration and exploration spaceflight. Establishing and maintaining a firmly grounded ethics framework for this inherently risky activity is essential to guide NASA's decisions today and to create a strong foundation for decisions about future challenges and opportunities.

REFERENCES

ALI (American Law Institute). 1979. Restatement of the law second, torts. Vol. 4. Philadelphia, PA: ALI.

Behnken, R., M. Barratt, S. Walker, and P. Whitson. 2013. *Presentation to the Institute of Medicine, Ethics Principles and Guidelines for Health Standards for Long Duration and Exploration Spaceflights: Astronaut Office*. PowerPoint presented at the second meeting of the IOM Committee on Ethics Principles and Guidelines for Health Standards for Long Duration and Exploration Spaceflights. Washington, DC, July 25. http://www.iom.edu/~/media/Files/Activity%20Files/Research/HealthStandardsSpaceflight/2013-JUL-25/Panel%202%20Astronaut%20Corp%20Final%20IOM_presentation final2.pdf (accessed November 8, 2013).

Boyd, A. D., P. R. Saxman, D. A. Hunscher, K. A. Smith, T. D. Morris, M. Kaston, F. Bayoff, B. Rogers, P. Hayes, N. Rajeev, E. Kline-Rogers, K. Eagle, D. Clauw, J. F. Greden, L. A. Green, and B. D. Athey. 2009. The University of Michigan honest broker: A web-based service for clinical and translational research and practice. *Journal of the American Medical Informatics Association* 16(6):784-791.

CIOMS (Council for International Organizations of Medical Sciences). 2002. *International ethical guidelines for biomedical research involving human subjects*. http://www.cioms.ch/publications/layout_guide2002.pdf (accessed March 10, 2014).

Emanuel, E. J., D. Wendler, and C. Grady. 2000. What makes clinical research ethical? *JAMA* 311(9):2701-2711.

Faden, R. R., and T. L. Beauchamp. 1986. *A history and theory of informed consent*. New York: Oxford University Press.

Faden, R. R., N. E. Kass, S. N. Goodman, P. Pronovost, S. Tunis, and T. L. Beauchamp. 2013. An ethics framework for a learning health care system: A departure from traditional research ethics and clinical ethics. *Hastings Center Report* 43:S16-S27.

Farber, D. A. 2011. Uncertainty. *The Georgetown Law Journal* 99:901-959.

Fletcher, C. E. 1988. Sophisticated investors under the federal securities laws. *Duke Law Journal* 1988(6):1081-1154.

Gostin, L. O. 1995. Informed consent, cultural sensitivity, and respect for persons. *JAMA* 274(10):844-845.

HHS (Department of Health and Human Services). 1993. Chapter 1A: Jurisdiction of the Institutional Review Board. In: *Institutional Review Board Guidebook*. Washington, DC: HHS.

IOM (Institute of Medicine). 2001. *Safe passage: Astronaut care for exploration missions*. Washington, DC: National Academy Press.

IOM. 2007. *The learning healthcare system: Workshop summary (IOM Roundtable on Evidence-Based Medicine)*. Washington, DC: The National Academies Press.

IOM. 2014. *Review of NASA's evidence reports on human health risks: 2013 letter report*. Washington, DC: The National Academies Press.

Kass, N. E., R. R. Faden, S. N. Goodman, P. Pronovost, S. Tunis, and T. L. Beauchamp. 2013. The research-treatment distinction: A problematic approach for determining which activities should have ethical oversight. *Hastings Center Report* 43:S4-S15.

Mazur, D. J. 2003. Influence of the law on risk and informed consent. *British Medical Journal* 327(7414):731-734.

NASA (National Aeronautics and Space Administration). 2007. *NASA space flight human system standard. Volume 1: Crew health*. NASA-STD-3001. https://standards.nasa.gov/documents/detail/3315622 (accessed December 4, 2013).

NASA. 2008. *Protection of human research subjects.* NASA Procedural Requirements 7100.1. http://nodis3.gsfc.nasa.gov/displayDir.cfm?t=NPR&c=7100&s=1 (accessed December 5, 2013).

NASA. 2012. *NASA space flight program and project management handbook and the standing review board handbook.* NASA Procedural Requirements 7120.5E. http://nodis3.gsfc.nasa.gov/displayDir.cfm?t=NPR&c=7120&s=5E. (accessed February 25, 2014).

NASA. 2013. *NASA policy for safety and mission success.* http//nodis3.gsfc. http://nodis3.gsfc.nasa.gov/displayDir.cfm?t=NPD&c=8700&s=1E (accessed February 25, 2014).

NIDCR (National Institute of Dental and Craniofacial Research). 2014. *Data and Safety Monitoring Board (DSMB) guidelines.* http://www.nidcr.nih.gov/Research/ToolsforResearchers/Toolkit/DSMBGuidelines.htm (accessed January 15, 2014).

Schwartz, J. 2010. Fairness, utility, and market risk. *Oregon Law Review* 89(1):175-262.

Tauber, A. I. 2005. *Patient autonomy and the ethics of responsibility.* Cambridge, MA: MIT Press.

A

Meeting Agendas

COMMITTEE ON ETHICS PRINCIPLES AND GUIDELINES FOR HEALTH STANDARDS FOR LONG DURATION AND EXPLORATION SPACEFLIGHTS

National Academy of Sciences Building, Room 125
2101 Constitution Avenue, NW
Washington, DC 20001

OPEN SESSION – May 30, 2013

10:45 – 11:15 a.m. **Welcome and Introductions**
Carol Scott-Conner, Chair, IOM Standing Committee on Aerospace Medicine and the Medicine of Extreme Environments
Jeffrey Kahn, Chair, IOM Committee on Ethics Principles and Guidelines for Health Standards for Long Duration and Exploration Spaceflights

11:15 a.m. – 12:00 p.m. **Charge to the New Committee**
Richard Williams, Chief Health and Medical Officer, NASA
Jeffrey Davis, Director, Human Health and Performance Directorate, Johnson Space Center, NASA

12:00 – 1:00 p.m.	**Committee Discussion with NASA Staff**
	Lunch
1:00 – 1:30 p.m.	**NASA and the Importance of Risk** *Charles Bolden, Administrator, NASA*
1:30 – 3:15 p.m.	**Risk Acceptance and Health Standards**
1:30 – 1:50 p.m.	NASA Risk Acceptance – *George Gafka, Chief, Safety Officer, International Space Station Program, NASA*
1:50 – 2:10 p.m.	NASA Health Standards – *David Liskowsky, Director, Medical Policy and Ethics, NASA*
2:10 – 2:30 p.m.	Ethics of Standard Setting – *Paul R. Wolpe, Bioethicist, NASA*
2:30 – 3:15 p.m.	Discussion
3:15 – 3:30 p.m.	**Break**
3:30 – 5:00 p.m.	**Specific Health Issues**
3:30 – 4:15 p.m.	Radiation – *Francis A. Cucinotta, Chief Scientist, NASA Space Radiation*
4:15 – 5:00 p.m.	Ocular and Intracranial Pressure – *Dave Francisco, Program Integration Manager, Human Health and Performance, NASA* *Terrance Taddeo, Chief, Space and Clinical Operations Division, NASA*
5:00 – 5:30 p.m.	**Revisit the Statement of Task and Questions for NASA Staff**
5:30 p.m.	**Adjourn**

COMMITTEE ON ETHICS PRINCIPLES AND GUIDELINES FOR HEALTH STANDARDS FOR LONG DURATION AND EXPLORATION SPACEFLIGHTS

National Academy of Sciences Building, Room 120
2101 Constitution Avenue, NW
Washington, DC 20001

OPEN SESSION – July 25, 2013

8:00 – 8:05 a.m. **Welcome**
Jeffrey Kahn, Committee Chair

8:05 – 8:30 a.m. **Decision Making in the Context of Uncertainty**
Harvey V. Fineberg, Institute of Medicine

8:30 – 9:25 a.m. **Panel 1: Risk Acceptance and Risk Management at NASA**
Facilitator: Carol Scott-Conner, Committee Member

8:30 – 8:35 a.m. Panel Introductions

8:35 – 8:50 a.m. Evolution of Risk Management at NASA–*Homayoon Dezfuli, NASA*

8:50 – 9:05 a.m. External Regulatory Framework for NASA Health Standards–*David Liskowsky, NASA*

9:05 – 9:25 a.m. Discussion with the Committee

Panel Questions:

- *How has risk management evolved at NASA over the years?*
- *What is the external regulatory framework under which NASA develops health standards?*

- *How do OSHA, NRC, and other standards and regulations impact NASA standards and decision making?*
- *Are risk considerations different for long duration and exploration missions? Does the risk management approach change?*

9:25 – 10:35 a.m. **Panel 2: Perspective from the Astronaut Corps**
Facilitator: Bonnie Dunbar, Committee Member

9:25 – 9:30 a.m. Panel Introductions

9:30 – 10:10 a.m. Perspectives on Risk Management and Risk Acceptance
Peggy Whitson, NASA (via video conference)
Mike Barratt, NASA
Shannon Walker, NASA
Robert Behnken, NASA

10:10 – 10:35 a.m. Discussion with the Committee

Panel Questions:

- *Are risk considerations different for long duration and exploration missions? Does the risk management approach change?*
- *In light of the unknowns and uncertainties, how much risk can be taken? Who makes those decisions? What data and considerations go into those decisions?*
- *What input do astronauts have into decisions regarding risk and health standards? What are the processes for that input?*

10:35 – 10:50 a.m. **Break**

10:50 – 12:20 p.m. **Panel 3: Ethics Framework for Considering Health Standards for Long Duration and Exploration Spaceflights**
Facilitator: Jonathan Kimmelman, Committee Member

10:50 – 10:55 a.m. Panel Introductions

10:55 – 11:25 a.m. Perspectives on an Ethics Framework–
Laurie Zoloth, Northwestern University, and Paul Root Wolpe, Emory University/NASA (via video conference)

11:25 – 11:40 a.m. Ethics, Risk, and Individual Differences–
Alta Charo, University of Wisconsin–Madison (via video conference)

11:40 – 11:55 a.m. Informed Consent in High-Uncertainty Contexts–
David Wendler, National Institutes of Health

11:55 – 12:20 p.m. Discussion with the Committee

Panel Questions:

- *What ethical principles and considerations should guide establishment and application of health standards concerning uncertain health effects of long duration and exploration spaceflight? What ethical framework can guide development and interpretation of such standards?*
- *How does the nature of activities (e.g., exploration, directed mission, routine tasks) alter ethical evaluation of risk exposure in spaceflight? How should spaceflight agencies weigh the risks and benefits? Are there absolute limits on acceptable risk?*

- *How can health standards be established in light of large uncertainties? Should current health standards be broadened for exploration missions? If so, what are the ethical considerations regarding how much risk can be taken?*
- *What are the ethical considerations about the role of individual susceptibilities setting and applying health standards (e.g., age, gender, genetics)?*
- *What issues regarding informed consent should the committee consider?*

12:20 – 1:15 p.m. **Working Lunch**

1:15 – 2:30 p.m. **Panel 4: Decision Making and Uncertain Risk: Learning from Other Occupations**
Facilitator: Larry Palinkas, Committee Member

1:15 – 1:20 p.m. Panel Introductions

1:20 – 1:35 p.m. Professional Sports and High Degrees of Physical Risk–*Sean Sansiveri, NFL Players Union (via video conference)*

1:35 – 1:50 p.m. Diving in Extreme Environments: The Scientific Diving Experience–*Michael A. Lang, American Academy of Underwater Sciences (via video conference)*

1:50 – 2:05 p.m. Psychological and Physical Health Risks at the Circumpoles–*Peter Suedfeld, University of British Columbia (via video conference)*

2:05 – 2:30 p.m. Discussion with the Committee

Panel Questions:

- *What is the purpose/benefit of activities in your field?*
- *In general, what are the major known and uncertain risks in your field?*

- *How/why do benefits outweigh these risks, justifying societal and individual acceptance of identified risks?*
- *Are limits placed on individual acceptance of risk in your field? In other words, what criteria or health standards govern whether individuals are allowed to expose themselves to known or uncertain risks? Briefly describe the major standards/limitations. Do these standards/limitations change in the context of uncertain risk?*
- *How are individuals informed about risks and benefits, and what are the processes for informed consent/decision making?*
- *How have the risks changed in your field based on new evidence, the ability to detect risk, or improved protections (e.g., technological or engineering)? Give an example of how health standards/criteria have changed in light of new evidence about risk.*
- *What lessons can NASA draw from your experiences?*

2:30 – 2:45 p.m. **Break**

2:45 – 4:10 p.m. **Panel 5: Issues in Health Standard Implementation and Enforcement**
Facilitator: Michael Silverstein, Committee Member

2:45 – 3:10 p.m. Perspective on OSHA Standards–*John Mendeloff, University of Pittsburgh*

3:10 – 3:25 p.m. FAA Medical Standards–*Fred Tilton, Federal Aviation Administration*

3:25 – 3:45 p.m. Ethics, Risk, and the Problems with Trimming Radiation-Dose Data–*Kristin Shrader-Frechette,*

University of Notre Dame

3:45 – 4:10 p.m. Discussion with the Committee

Panel Questions:

- *What laws and regulations govern health standard implementation and enforcement in your agency? What are these standards?*
- *How are risks and benefits weighed in selection of health standards? In other words, what tips the balance between acceptable and unacceptable risk?*
- *Are standards universally applied or tailored to individuals, and what are the associated ethical issues?*
- *Can existing health standards be modified or waived? If so, under what circumstances can risk exposures exceed set standards and what is the rationale for allowing this?*
- *Are there examples of how health standard selection or enforcement criteria have evolved in response to new evidence or shifts in culture?*
- *What lessons can NASA draw from your experiences?*

4:10 – 4:45 p.m. **Public Testimony – Registered Speakers**
Moderator: Jeff Kahn, Committee Member
(3 minutes per speaker)

4:45 p.m. **Public Session Adjourn**

B

Committee Biographical Sketches

Jeffrey Kahn, Ph.D., M.P.H. (*Chair*), is the inaugural Robert Henry Levi and Ryda Hecht Levi Professor of Bioethics and Public Policy at the Johns Hopkins University Berman Institute of Bioethics. Prior to joining the faculty at Johns Hopkins in 2011, Dr. Kahn was director of the Center for Bioethics and the Maas Family Endowed Chair in Bioethics at the University of Minnesota. Earlier in his career, Dr. Kahn was director of the Graduate Program in Bioethics and assistant professor of bioethics at the Medical College of Wisconsin, and associate director of the White House Advisory Committee on Human Radiation Experiments. Dr. Kahn works in a variety of areas of bioethics, exploring the intersection of ethics and public health policy, including research ethics, ethics and genetics, and ethical issues in public health. He has served on many advisory panels, including the National Institutes of Health (NIH), the Centers for Disease Control and Prevention (CDC), the Institute of Medicine (IOM), among others, and speaks nationally and internationally on a range of bioethics topics. He has published 3 books and more than 125 articles in the bioethics and medical literature. From 1998 to 2002, he wrote the biweekly column "Ethics Matters" for CNN.com. Dr. Kahn earned his B.A. in Microbiology from the University of California, Los Angeles, his M.P.H. from Johns Hopkins University, and his Ph.D. in Philosophy/Bioethics from Georgetown University.

Nancy Conrad is the founder and chair of The Conrad Foundation, a nonprofit organization focused on energizing and engaging students in science and technology through unique entrepreneurial opportunities. The foundation was founded as a legacy to Apollo 12 astronaut and entrepreneur, Charles "Pete" Conrad. She is currently the education partner for the SonUS Solar-Sail-Sun Jammer project as well the Inspiration Mars project, and is on the advisory board for the B612 Foundation. Ms. Conrad is a founding partner of the Global Patient Safety Team Advisory Board. She serves on the Social Enterprise Governing Board at Columbia University. Ms. Conrad is a member of the President's Circle of the National Academy of Sciences. As a leader in transformative education, she has testified before the U.S. House of Representatives Committee on Science, Space, and Technology on the use of partnerships and mentorship to improve science, technology, engineering, and mathematics (STEM) education. Ms. Conrad is affiliated with a number of organizations that emphasize STEM education, including the STEM Education Coalition and STEMconnector.

Peter F. Demitry, M.D., M.P.H., has been president of 4-D Enterprises, LLC, an executive consulting service, since 2006. He served on the Air Force Scientific Advisory Board's quick-look study of the F-22's oxygen system. Prior to that, he led a Traumatic Brain Injury (TBI) Consortium, an industry-based group focused on transitioning technologies such as sensors, biomarkers, diagnostics, and therapeutics for TBI. Dr. Demitry previously served as the first Assistant Air Force Surgeon General for Modernization from 2002 to 2006, during which time he also served as the Chief Information Officer and Chief Technology Officer for the Air Force Medical Service, a $6 billion global health care system. In that role he was responsible for transitioning all medical technologies into the U.S. Air Force. He retired from the Air Force in 2006 after serving more than 28 years on active duty. Dr. Demitry received his undergraduate degree from the U.S. Air Force Academy, his M.D. from the Uniformed Services University of the Health Sciences, and a master's degree in public health from the Harvard School of Public Health. He completed his occupational medicine residency at Harvard University and became certified in occupational and environmental medicine. Dr. Demitry is a certified Air Force test pilot and is the only test pilot physician in Air Force history. He is a Fellow of the Aerospace Medical Association and earned membership within the Society of Experimental Test Pilots. He flew the F-4, A-10, F-16, and F-15 in flying assignments and was the director and chief test pilot of the Advanced Fighter Technology F-16 Integration program at Edwards Air Force Base. Among his awards, he has received the Meritorious Service Medal, the Aerial Achievement Medal, and the Legion of Merit Award.

Bonnie J. Dunbar, Ph.D., is a former astronaut for the National Aeronautics and Space Administration (NASA). She retired from NASA in 2005. She then served as president and chief executive officer (CEO) of The Museum of Flight until 2010. She is now director of higher education and strategic workforce planning for the Boeing Company. Following graduation in 1971 from the University of Washington, she worked for Boeing Computer Services as a systems analyst. Dr. Dunbar has served as an adjunct assistant professor in mechanical engineering at the University of Houston. Dr. Dunbar is a private pilot with more than 200 hours in single-engine land aircraft, has logged more than 700 hours flying time in T-38 jets as a backseater, and has more than 100 hours as a co-pilot in a Cessna Citation jet. She accepted a position as a payload officer/flight controller at the Lyndon B. Johnson Space Center in 1978. She served as a guidance and navigation officer/flight controller for the Skylab reentry mission in 1979 and was subsequently designated project officer/payload officer for the integration of several Space Shuttle payloads. Dr. Dunbar became a NASA astronaut in 1981. Her technical assignments have included assisting in the verification of Shuttle flight software at the Shuttle Avionics Integration Laboratory (SAIL), serving as a member of the Flight Crew Equipment Control Board, participating as a member of the Astronaut Office Science Support Group, and supporting operational development of the remote manipulator system (RMS). In 1993, Dr. Dunbar served as Deputy Associate Administrator, Office of Life and Microgravity Sciences, NASA Headquarters. She spent 13 months training as a back-up crew member for a 3-month flight on the Russian Space Station, Mir. From 1995 to 1996, she was detailed to the NASA Johnson Space Center Mission Operations Directorate as assistant director. She was responsible for chairing the International Space Station Training Readiness Reviews, and facilitating Russian/American operations and training strategies.

Barbara J. Evans, B.S.E.E., M.S., Ph.D., J.D., LL.M., is a professor of law and George Butler Research Professor and is director of the Center on Biotechnology & Law at the University of Houston Law Center. She was a Greenwall Foundation Faculty Scholar in Bioethics from 2010 to 2013. Her research interests include governance and privacy issues with large health information networks and regulatory uses of scientific evidence to assess and manage health risks in various contexts, including postmarketing drug safety oversight, regulatory approval of novel medical products, environmental regulation of agricultural biotechnologies, and oversight of health care quality and patient safety. Earlier in her career, she was a partner in the international regulatory practice of a large New York law firm and subsequently advised clients on U.S. privacy, research, and medical device regulatory matters. Prior to joining the University of Houston Law Center, she was a research professor of medicine and director of the Program in Pharmacogenomics, Ethics, and Public Policy at the Indiana University School of Medicine/Center for Bioethics. She holds a B.S.E.E. from the University of Texas at Austin; an M.S. and a Ph.D. from Stanford University; a J.D. from Yale Law School; and an LL.M. in health law from the University of Houston Law Center. She also completed a postdoctoral fellowship in clinical ethics at the M.D. Anderson Cancer Center.

Bernard A. Harris, Jr., M.D., M.B.A., is CEO and managing partner of Vesalius Ventures, Inc. Dr. Harris was at NASA for 10 years, where he conducted research in musculoskeletal physiology and disuse osteoporosis. He conducted clinical investigations of space adaptation and developed in-flight medical devices to extend astronaut stays in space. Selected into the Astronaut Corps in 1990, Dr. Harris was a mission specialist on the Space Shuttle Columbia STS-55/Spacelab D-2 in 1993. As payload commander on Space Shuttle Discovery STS-63 in 1995, he served on the first flight of the joint Russian–American Space Program, becoming the "First African American to walk in space." A veteran astronaut for more than 20 years, he has logged more than 438 hours and traveled more than 7.2 million miles in space. He has had more than 20 years of business experience in various leadership roles, such as CEO, president, and chief medical officer of some of the leading companies and organizations in the nation. In addition, he is on the boards of the Houston Technology Center, National Space Biomedical Research Institute, Board of Scientific Counselors, HealthConnect, and National Math and Science Initiative. Dr. Harris is also the founder of the Harris Foundation, a nonprofit organization that supports math/science education and crime prevention programs for America's youth. He earned a B.S. in biology from the University of Houston, a master of medical science from the University of Texas Medical Branch at Galveston, an M.B.A. from the University of Houston, and an M.D. from Texas Tech University School of Medicine. He completed a residency in internal medicine at the Mayo Clinic, a National Research Council Fellowship in Endocrinology at the NASA Ames Research Center, and trained as a flight surgeon at the Aerospace School of Medicine, Brooks Air Force Base. He is also a licensed private pilot and certified scuba diver. Dr. Harris is the recipient of numerous awards, including honorary doctorates from Stony Brook University, Morehouse School of Medicine, New Jersey Institute of Technology, and University of Hartford; a NASA Space Flight Medal; a NASA Award of Merit; and the 2000 Horatio Alger Award. He is also a Fellow of the American College of Physicians. He is the author of *Dream Walker: A Journey of Achievement and Inspiration* (Greenleaf Book Group Press, 2010).

David G. Hoel, Ph.D., is a distinguished university professor in the Department of Medicine at the Medical University of South Carolina. He also is a principal scientist at Exponent, Inc. Dr. Hoel was at the NIH National Institute of Environmental Health Sciences for more than 20 years as director of the Division of Environmental Risk Assessment, which developed methods for quantitatively estimating health risks from low-dose chemical exposures. He has particular interests in estimating the health effects of radiation exposures and spent a total of 3 years working at the Radiation Effects Research Foundation in Hiroshima, Japan, as a program director. His current research is focused on low-dose adverse health effects of gamma, neutron, and alpha radiation as well as plutonium in particular. His research support has included a 5-year project for NASA on the analysis of the potential health risks from high linear energy transfer radiation. International committees on which he has served include a radiation exposure advisory committee for the International Atomic Energy Agency and the World Health Organization's International Agency for Research on Cancer. He is an IOM member and a Fellow of the American Association for the Advancement of Science (AAAS). He earned a B.A. in mathematics from the University of California, Berkeley, and a Ph.D. in mathematical statistics from the University of North Carolina at Chapel Hill. He completed postdoctoral training in preventive medicine from Stanford University. Dr. Hoel has served on the National Research Council's (NRC's) Committee on Evaluation of Radiation Shielding for Space Exploration, Committee on the Evaluation of Space Radiation Cancer Risk Models, and the Nuclear and Radiation Studies Board.

Jonathan Kimmelman, Ph.D., holds a doctorate in molecular biophysics and biochemistry from Yale University and is associate professor in biomedical ethics at McGill University, with a cross-appointment in experimental medicine. His research centers on the ethics of translational clinical research. He leads several funded projects investigating risk-benefit across the research trajectory, and directs the Studies for Translation, Ethics, and Medicine (STREAM) Group. Major publications have appeared in *Science*, *Lancet*, *BMJ*, *PLoS Medicine*, and *Hastings Center Report*. His book, *Gene Transfer and the Ethics of First-in-Human Experiments* (Cambridge Press, 2010), is the first full-length analysis of the ethics of translational clinical research and has been described as "set[ting] a new standard for bioethical scholarship that is at once scientifically well-grounded, politically astute, philosophically original, and a pleasure to read." Dr. Kimmelman was the winner of the 2006 Maud Menten New Investigator Prize (Institute of Genetics) and received a Canadian Institutes of Health Research New Investigator Salary Award in 2008. He has served in numerous advisory capacities, including ethics committee chairs for the American Society of Gene and Cell Therapy (2008-2010) and the International Society of Stem Cell Research (since 2013). He is a member of the National Heart, Lung, and Blood Institute Gene and Cell Therapy Data Safety Monitoring Board.

Anna C. Mastroianni, J.D., M.P.H., is professor of law at the University of Washington School of Law. She has additional faculty appointments in the university's School of Public Health and School of Medicine. Prior to her academic career, she held a number of legal and federal policy positions in Washington, DC, including associate director of President Clinton's Advisory Committee on Human Radiation Experiments and study director of the IOM of the National Academy of Sciences (NAS). She also practiced health law with two private law firms in Washington, DC. She has served on a number of committees that advise the U.S. government and other entities, including the NRC's Committee on Institutional Review Boards, Social Science and Surveys; the IOM's Committee on the Review of the National Immunization Program's Research Procedures and Data Sharing Program; and the NIH Recombinant DNA Advisory Committee. In addition, she has been nationally recognized for her contributions to health policy, law, and bioethics as a Fellow of AAAS. Her publications include six books and numerous peer-reviewed articles on law, medicine, and bioethics, with a special emphasis on the legal and ethical challenges in public health, research with human subjects, and assisted reproductive technologies. Professor Mastroianni is a graduate of the University of Pennsylvania's School of Law (J.D.), The Wharton School (B.S. in economics), College of Arts and Sciences (B.A. in Spanish and Portuguese), and the University of Washington School of Public Health (M.P.H. in health services).

Lawrence Palinkas, Ph.D., is the Albert G. and Frances Lomas Feldman Professor of Social Policy and Health in the School of Social Work at the University of Southern California (USC). He also holds appointments in the Departments of Anthropology and Preventive Medicine at USC and serves as adjunct professor of medicine and family and preventive medicine at the University of California, San Diego. A medical anthropologist, his primary areas of expertise lie within behavioral health, preventive medicine, crosscultural medicine, and health services research. Dr. Palinkas is particularly interested in health disparities, implementation science, community-based participatory research, and the sociocultural and environmental determinants of health and health-related behavior, with a focus on disease prevention and health promotion. Current research encompasses mental health services, immigrant health, and global health. Specific projects explore the mental health needs of individuals in extreme and unusual environments and communities impacted by natural and manmade disasters; cultural explanatory models of mental illness and service use; evaluation of academic–community research practice partnerships; and the dissemination and implementation of evidence-based practices for delivery of mental health services to children, adolescents, and underserved populations. He also provides expertise to students and colleagues in the use of qualitative and mixed research methods. Among his scholarly achievements are the Antarctic Service Medal by the National Science Foundation and the U.S. Navy; deputy chief officer of the Life Sciences Standing Scientific Committee on Antarctic Research; chair of the National Space Biomedical Research Institute's External Advisory Council; and membership on committees of the NRC, NAS, and the IOM. Dr. Palinkas is an elected Fellow of the American Anthropological Association and Society for Applied Anthropology.

Carol E. H. Scott-Conner, M.D., Ph.D., M.B.A., is a professor in the department of surgery at the University of Iowa, Iowa City. Dr. Scott-Conner received her undergraduate training in electrical engineering from the Massachusetts Institute of Technology, and worked as an engineer before attending medical school at New York University (NYU). She received her M.D. from NYU, where she completed a residency in surgery. After leaving NYU, she joined the faculty at Marshall University, and then moved to the University of Mississippi. During her tenure there she earned a Ph.D. in anatomy from the University of Kentucky, and an M.B.A. In 1995, she became professor and head of surgery at the University of Iowa. Dr. Scott-Conner has been active on 22 editorial boards, and has authored more than 200 original papers, abstracts, reviews, and book chapters. She is certified by the National Board of Medical Examiners and the American Board of Surgery and has a Certification of Added Qualifications in Surgical Critical Care. Dr. Scott-Conner has served on a number of IOM committees and chairs the IOM Standing Committee on Aerospace Medicine and the Medicine of Extreme Environments.

Michael A. Silverstein, M.D., M.P.H., recently retired from his post as the assistant director for industrial safety and health in the Washington State Department of Labor and Industries after directing the state's Occupational Safety and Health Administration (OSHA) program for 10 years. Dr. Silverstein is currently a clinical professor of environmental and occupational health at the University of Washington School of Public Health. Dr. Silverstein also held positions in the Washington State Department of Health as state health officer and epidemiologist and spent 2 years in Washington, DC, as director of policy for OSHA. For 15 years before this, he was assistant director for occupational health and safety with the United Automobile Workers Union in Detroit. Dr. Silverstein has practiced family and occupational medicine in Michigan and California. He has degrees from Harvard University, Stanford Medical School, and the University of Michigan School of Public Health. He is board certified as a specialist in occupational medicine. Dr. Silverstein has been an active member of several professional associations, including the American Public Health Association, where he served as chair of the occupational safety and health section, and the American College of Occupational and Environmental Medicine, where he was a member of the Ethics Committee. He spent 2 years as chair of the National Advisory Committee on Occupational Safety and Health. He has authored numerous scientific research and policy articles, including recent publications on the regulatory process, the aging workforce, the future of OSHA, and asbestos cancer risk assessment. Dr. Silverstein has served on several IOM panels, including the Committee on Aerospace Medicine and Medicine of Extreme Environments, the Committee on NASA's Research on Human Health Risks, the Committee to Review NASA's Space Flight Standards, the Committee to Review the National Institute for Occupational Safety and Health (NIOSH) Hearing Loss Research program, the Committee on the Health and Safety Needs of Older Workers, the Committee on Health and Safety Implications of Child Labor, and the Transportation Research Board Committee on Offshore Windfarm Worker Safety.

Ronald E. Turner, Ph.D., is a Distinguished Analyst at Analytic Services Inc. He has more than 25 years of experience, including expertise in space physics, life science systems, and space policy. He is the senior science advisor to the NASA Innovative Advanced Concepts Program. He is also an internationally recognized expert in radiation risk management for astronauts, particularly in response to solar storms, and he has frequently been an invited speaker to describe radiation risk management strategies. He has participated in several NRC studies of radiation risk management for exploration missions, both as panelist and as a reviewer, including Space Radiation Hazards and the Vision for Space Exploration, Managing Space Radiation Risks in the New Era of Space Exploration, Technical Evaluation of the NASA Model for Cancer Risk (reviewer), and NASA Space Technology Roadmaps and Priorities, Human Health and Exploration Systems (panelist). He was on the Advisory Council to the National Space Biomedical Research Institute Center for Acute Radiation Research. He led a NASA Office of the Chief Engineer study to understand NASA's requirements for operational space weather support. He is a member of the International Academy of Astronautics, and also belongs to the American Geophysical Union and the American Institute of Aeronautics and Astronautics. He has a Ph.D. in nuclear/particle physics from the Ohio State University and M.S. and B.S. from the University of Florida.

R. Leonard Vance, Ph.D., J.D., is an associate professor in the Center for Environmental Studies at Virginia Commonwealth University. His research interests include exposure monitoring of environmental and occupational contaminants in workers, law and regulation of chemicals, applied industrial hygiene, asbestos and lead control, and engineering controls for toxic substances. From 1982 to 1986, Dr. Vance served as the director of health standards for OSHA. From 1976 to 1982, he served as the Assistant Attorney General of Virginia. Dr. Vance received a Ph.D. in inorganic chemistry from the University of Virginia and his law degree from the University of Richmond. He is a licensed professional engineer and is board certified in industrial hygiene, safety, and hazardous materials management.

Gregory R. Wagner, M.D., is senior advisor to the director of NIOSH and adjunct professor of environmental health at Harvard School of Public Health. He recently returned to NIOSH after serving as Deputy Assistant Secretary of Labor for Mine Safety and Health. While at NIOSH, Dr. Wagner directed the Division of Respiratory Disease Studies, served as Associate Director for Mining, and coordinated a public research priority-setting process resulting in the National Occupational Research Agenda that guided U.S. research for a decade. He served on the Ethics Committee that rewrote the code of ethics for the American College of Occupational and Environmental Medicine, chaired the Ethics Committee of the American Thoracic Society, and also helped develop the original Code of Ethics for the Association of Occupational and Environmental Clinics. He previously served as a medical officer in the National Health Service Corps of the U.S. Public Health Service in rural West Virginia. A graduate of Harvard College and Albert Einstein College of Medicine, Dr. Wagner has both taught and practiced internal and public health/occupational medicine, and is board certified in both fields.